Reforming Science

Reforming Science
Beyond Belief

Brian Ridley

imprint-academic.com

Copyright © Brian Ridley, 2010

The moral rights of the author have been asserted
No part of this publication may be reproduced in any form
without permission, except for the quotation of brief passages
in criticism and discussion.

Published in the UK by Imprint Academic
PO Box 200, Exeter EX5 5YX, UK

Published in the USA by Imprint Academic
Philosophy Documentation Center
PO Box 7147, Charlottesville, VA 22906-7147, USA

ISBN 9 781845 40194 8

A CIP catalogue record for this book is available from the
British Library and US Library of Congress

Contents

Introduction: Beyond Hubrisci 1
1 The Soul, No Matter, Never Mind 7
2 A Short History of Animism 15
3 A Magical World 29
4 The New Science 43
5 For the Good of Humanity 57
6 The New Philosophy 69
7 Panpsychism 85
8 The Mind-Body Interaction 97
9 A Strange New Science 113
10 Towards a Mathematical Theology 127
11 Origins 143
12 The Big Bang Story 155
13 Meta-Cosmology 171
14 Beyond Belief 181
Sources of Quotations 191
Short Bibliography 193
Index 195

Introduction

There is an urgent need for a reformation of science: science has become middle-aged and its age is beginning to show. It no longer has the exuberance and self-confidence of youth, no longer has the zip that astonished the world with Newtonian gravity, Darwin's evolution, the quantum and relativity. In short, science is no longer gay, in the Nietzschean sense.

The fact is, it has become a mite touchy. Its well-deserved pride at its achievements, its satisfaction with the technological exploitation of those achievements which have benefited humanity, seem to have given rise to a certain pomposity in some of its established figures. These are like newly ennobled citizens, conscious of their importance in the natural order of things, who have adopted an aura of gravity of the sort reminiscent of what in *Tristram Shandy* is called 'a mysterious carriage of the body to cover the defects of the mind', and whose achievements are not to be questioned but, rather, lauded. Science has developed a public image of unchallengeable authority. There is a whiff of Church, a sense of establishment and infallibility, of powerful bishops, of jealously guarded reputations, of the dangers of heresy. The credo 'there is no truth but scientific truth' undermining the youthful romanticism of the search for Truth by whatever means. How foolish to question the fact that Darwin has solved the mystery of life, or that the problem of creation is solved by the Big Bang theory! But the whole Truth may lie elsewhere. In the current scientific climate, that thought is heretical. Which is why reform is overdue; which is why it is time for scientists to rediscover their intellectual consciences and cease to depend solely on 'received wisdom', time to escape from what might be called the hubrisci—the hubris and dogma—of much of modern science. It is time to question some of this 'wisdom', in particular, the belief that what isn't science, isn't knowledge. It is time to remind scientists they must be beyond belief—in the Humean sceptical sense. Science practised otherwise is what I have called hubrisci. Unfortunately, this is what many strands of science have become.

The first element of reformation is the deepest, namely, the reformulation of the meaning of matter and materialism. Conventional materialism is the doctrine that matter is all there is and that non-material entities such as the mind, the soul, God, do not exist. Everything functions according to the laws of physics, chemistry and biology, and only the methods of science can provide those laws. Now it is certainly the case that science cannot possibly function without the basic and deep assumption that all there is in the universe is matter subject to the methods of science, so it is inevitable that materialism must be its working creed. But materialism has no justification or credibility beyond that. The caricature of the scientist as a closed-minded materialist, reducing all human passions to the activity of neurons in the brain is only too well-known. No one doubts the value of the scientific method, but it is not easy to understand a mind that dismisses what is most immediately experienced—pain, excitement, love—as somehow unreal. Reformation must first restore the reality of mind as inhabiting Nature just as evidently as matter. The philosophic monism that takes matter to be all there is, frankly, is too attractive to give up, but this necessitates a fundamental reappraisal of what is meant by matter. For a start, it must include mind as one of its attributes, and matter as one of the attributes of mind. In other words, a reformulated materialism must espouse some form of panpsychism, the belief that mind in some degree is a universal attribute of matter.

A reformed science must also acknowledge its limitations, particularly in its analysis of life and the mental phenomena of the brain. Darwinian evolution, DNA genetics, neurophysiology, all encounter limitations. Science cannot study anything that is unique — it always needs a population — so the study of our unique universe hits a limitation straight away, which requires that all cosmological models have to be looked at askance. Where limitation is encountered — the essence of gravity, of the electron, of consciousness, and much else — the response has to be a deep feeling of awe and wonder at these cosmic mysteries. Science should not neglect to teach that wonder. The magical richness of the world should not be submerged by technicalities. It is simply the case that we are able to describe how a force like gravity works without having the least idea of what gravity is. The same is true for electric and magnetic forces and for the powers that inhabit the nucleus of the atom. These famous forces of physics are the paradigms of the unknowable cosmic mysteries of the universe, the fundamental attributes of Nature that are simply given. In accepting this, science has no excuse for not considering mind, and even life itself, as belonging to the same category of natural unknowables. But it is a category that sits very uneasily in the dogma of conventional materialism. So much the worse for conventional materialism and the hubrisci it generates.

Science was, without doubt, right to elevate matter over spirit in its early days, given the baleful influence of the Church and prevalence of superstition. In his advice to those fledgling scientists, Sir Francis Bacon, philosopher and Lord Chancellor of England, no less, urged that God be kept out of the laboratory. Given that all scientific knowledge had to be derived from experiments, and given that experiments consisted in observing matter and its responses to pushes and pulls, it was certainly necessary that concepts outside of matter and its responses should not be introduced. The discovery of gravity by Sir Isaac Newton was, in the Baconian sense, heretical and felt as such by many at the time, including the philosopher René Descartes. But gravity, if seen as some sort of spirit, at least was not of the Church. Materialists were happy to accept it as a property of matter, and, later, to accept, in the same way, electromagnetism and the nuclear forces. If science was to continue successfully with its experimental methods, there was no other choice. But nothing further extraneous should interfere with the cool analytical study of Nature, not religion, not politics, not dogma, and, above all, not anything supernatural, the latter to be seen as an oxymoron, an incoherent conception. Quite right too! It is the emergence of scientific politics and scientific dogma in modern science—the hubrisci of modern science—that raises the need for reformation.

Early science was also quite right to ignore many of the teachings of the ancients—the Earth-centred universe, the perfection of circular motion, the four bodily humours of Hippocrates, and much more. The refutation by science of those parts of ancient wisdom that had to do with the natural world tended to lead scientists to consign the whole corpus of ancient thought to oblivion. But some survived—the materialist world of the Ionians, the purposeless world of the early atomists, the mathematical world of the Pythagoraeans—but the ideas of Plato and Aristotle concerning the soul, cosmic forms, teleology, seemed irrelevant. A reformed science will have to reconsider. It is a fine irony that science itself grew out of the 'irrelevant' ideas that informed natural magic, astrology and alchemy, practised by many in the Renaissance, a practise persisting through to the birth of science.

It is my feeling that science needs a good shaking to eliminate internal dogmas, to free editors of scientific journals from the clutches of the scientific establishment, to recover that feeling that anything goes, provided that ideas are rationally advocated with suggestions for realistic experiments. We already have plenty of rationally advocated theories — cosmic inflation, string theory, multiple universes, many-dimensions, replicating croutons — but they are without experimental evidence and few suggestions for realistic experiments. All of which triggers thoughts, possibly unfair, that a career in theology, now sadly uncool, might be satisfactorily replaced by a career in mathematical theology, which is seen as wonderfully cool.

In the chapters that follow, I want to survey elements of the intellectual world, some conventional others unconventional, that bear on the issue of the reformation of science. They will take us from questions about the soul, through animism and the magical world, to modern science itself and its humane motivations. They will lead to questions about mind and body and an approach to this suggested by quantum theory and, most unconventional of all, to panpsychism. It is also necessary to look in some detail at the rise of abstruse mathematics in modern physics, tending to a kind of mathematical theology. Finally, I look at the dogmas in modern science — neo-Darwinism, Big Bangism, cosmology — the dogmas, in fact, that have motivated this book. In doing so I have found it hard to resist noting a few stimulating heresies that have come to my attention, not, I may say, by reading the leading journals, but rather via a kind of samizdat. I have focused on ideas rather than anecdotes to support the case for reform, not that anecdotal evidence concerning difficulties encountered in publishing interesting ideas is in short supply, but I preferred to leave that to others.

For the reader's benefit I thought that I should come clean about my own opinions about all this:

1. There is a regrettable closed-mindedness exhibited by many scientists, among them certain influential elements of the scientific establishment.
2. Heresies — ideas rationally expressed that offend received wisdom — are a good thing.

3. Nothing exists in the universe that is not matter and force.
4. Mind and body are but different aspects of matter and force.
5. The soul is a shorthand expression for the organic holistic individuality of the living being. No question of survival after death.
6. I'm inclined to think that Life is a force, a creative one that animates matter. It joins gravity, the electroweak force and the strong nuclear force, as capable of explaining special properties of matter and, like each of them, is a given and utterly beyond science. It's an inclination that will persist until some laboratory creates a living, replicating lump of matter. Which means creating a living cell.

Chapter One

The Soul, No Matter, Never Mind

I would like to believe that I have a soul. I would like to believe that that soul is uniquely me. I would like to believe that it is immortal. But, come on, you can't believe all that these days! Why not? Because there is no scientific evidence for it. True, but so what? There is no scientific evidence of multiple universes, but a lot of people believe they exist. And, we will see in the chapters ahead, multiple universes are not the only things; a lot of people believe in the existence of universal deterministic laws of Nature; people believe in pre-historical mind-sets determining our behaviour today, some believe in the existence of cultural genes that replicate, physicists believe that the Higgs particle, the particle associated with conferring mass to other particles, exists—the list is impressively long. Determinism, the dogma that told us that we were nothing but predictable machines, died with the three-body problem long before quantum theory put the boot in, yet evolutionary psychology hasn't noticed. What the cultural gene is—just call me meme—is a reductionist mystery. The only belief that might get some empirical support is the Higgs particle, which could materialise in the CERN machine soon. I would be sorry if it didn't, given the persuasive arguments for its existence and the colossal expense of the Large Hadron Collider. But there are good persuasive arguments for multiple universes, as well as for the Higgs particle. Are there no equally persuasive arguments for the existence of the soul? Well, there are plenty of arguments dating from the Pre-Socratics, with refinements from Plato and Aristotle, plus inputs from the medieval scholastics and the hermetic intellectuals of the renaissance. But, today, science has overwhelmed them. Not a matter of logical confutation, more a matter of cultural overload and indifference that these old intuitions seem no longer relevant. These old intuitions still exist, but, somehow, they are no longer convincing.

So I should settle for being soulless, a thing of pure matter, in no sense immaterial, dust to dust and all that. But what is the matter? Does anyone know? They claimed that they did once: sublunary stuff subject to change, technically defined by the property of extension and by its deterministic laws of motion. In modern physics matter is not the cloddish thing it once was. Relativity has modified the concept of extension; classical determinism is not what it used to be; quantum determinism doesn't exist; but, still, matter is mindless. And, like Descartes, I have direct empirical evidence that mind exists, even if the soul doesn't. So I can't settle for being mindless! But now we have got to the old hairy problem: how does mind arise from matter, if matter is all there is?

But does the solution of that, if there is a solution, compensate for the loss of the soul. There is another aspect to the whole problem not usually considered in philosophically polite circles—not rational, not metaphysical, not scientific, maybe even frivolous—but one that considers that life would be a good deal more interesting were there immortal souls in each of us, and especially, were the whole world to be animate in some sense. Our horses, our pets, are certainly animate with minds of some sort that recognise us, often with some indication of pleasure. Wouldn't it be good to think that talking to our plants (which certainly happens anyway) may evoke some vegetable pleasure? The active recognition of other life, no matter how it may be scorned as pure anthropomorphism, is surely good. And in everyday life there is a good deal of naming of inanimate things, from those teddy bears in childhood to the cars of the adult. And how we berate, and even slap, machines that don't work and occasionally pat them on their metal skin when they do! The whole world is alive said the Ionian Greeks and, strangely enough, this hylozoism, albeit intermittently, is with us today. Wouldn't life be richer if it were true that mountains, lakes, forests, islands were psychically alive? Wordsworth knew his beloved lakes were, Sibelius knew his forests were, Hardy knew his Wessex was. A living universe is simply more interesting than an inanimate one. It would also be a magical one, as any follower of Hermes Trismagistus would tell you.

This aspect of the mind-body problem has more than a touch of Epicurean hedonism about it, not to mention the associated dubious atmosphere of sulphurous demonism. It will be surely repulsive to the inner life of the religious, in which only the soul counts. Having a soul is an immense responsibility, but like many serious concerns of the spirit it has tended to become submerged by more urgent topical demands largely unchallenged by a contemporary unbelieving priesthood. But once, and with some it is still the case, the care of the soul was central to life and to the life hereafter. In the middle ages, that inner life was all. Now, in the West, that inner life and the concepts and values that went with it have largely vanished, thanks, I suppose, to a simplistic and uncritical infusion of science and the secular values of the Enlightenment.

As a scientist I am deeply aware how misleading that infusion can be. On the one hand the Enlightenment's claim of the superiority of reason discounts the priority of intuition and passion, and on the other the brilliant success of science has given it a prestige far beyond its down-to-earth capabilities. Nevertheless, the idea of the supernatural sits uneasily in any world-view today. The old animistic belief in the soul leaving the body during sleep, and its wanderings accounting for dreams, has not completely vanished. The killing of wives and slaves on their master's death so that the shade of their master will continue to be well served is not very old history. Faustian demons are now encountered only in the theatre, but the spirits of the saints and angels and ancestors still exert powerful influences on some. If, as paid-up creatures of the Enlightenment, we disavow the supernatural, then we must at least consider both mind and matter as empirically observable facts of nature.

We know how to ascertain the properties of matter. We can prod it and probe it and kick it about using well-established scientific methods. That goes for the brain *as matter*, but those methods are clearly useless for the mind. As far as the enquiry into the nature of mind-matter is to be carried out, a dualistic approach is inevitable. This is not to say that we have to accept a dualism of the Descartesian kind — dual substances — but rather merely an operational dualism, each with its own language. Focus on matter and we will get a description characteristic of the sciences; focus on the mind and we will get a description characteristic of the humanities. In the one, mind is the epiphenomenon of the body; in the other, body is the epiphenomenon of the mind. In metaphysics, one leans towards materialism, the other to idealism, but metaphysics can be dispensed with in practice. In the day-to-day exploration that the curious make of these matters in their laboratories, in their studies, in their easy chairs, operational dualism is the only possible assumption to make, outside of dogmatism.

But none of this solves the problem of the mind-body interaction. Spinoza, who was the first to see mind and matter as the dual aspects of a universal substance, namely, God, saw no sense in the concept of them interacting, since they were one. Moreover, if they were not one, they could not possibly interact: 'If two things have nothing in common with one another, the one cannot be the cause of the other.' Spinoza's dual aspect idea needs his full-blown pantheism to support it. But we can retain the dual aspect of mind and matter without the pantheism by noting that different aspects do not necessarily imply a lack of common properties. Here an analogy with quantum theory is illuminating. A submicroscopic particle like the electron has a dual aspect: in one situation it is a wave, in another situation it is a particle. Nature reveals this dual aspect at the most fundamental level. Both wave and particle aspects are essential attributes of this thing that is the electron, and indeed of all elementary particles. To speak of a cause-and-effect type of interaction between the wave and the particle is meaningless. Focus on the wave-like property of the electron and it is a wave; focus on the particle-like property and it is a particle. Wave and particle are complementary aspects of a reality that is a single entity. The analogy with the mind-matter dual aspect of the brain is striking. It suggests that to speak of a cause-and-effect interaction between mind and brain-matter is meaningless: they are the same entity. Moreover, the complementary nature suggests that focus on one drives out the other. Enquire about neural paths in the brain, and the mind collapses; enquire about thoughts and those neural pathways become irrelevant. Mind and organic neural matter are complementary aspects of the brain of each of us.

Extending, by analogy, the quantum-theoretic principle of complementarity to the brains of humans and animals may be productive, but its extension to all living things is more problematical. We would need the belief that every sentient being with a neural system will thereby possess a mind, be its manifestation only of instinctive, purposeful behaviour. Going further and assigning mind to the daisy is an even bigger step, and the biggest step of all, assigning mind to inanimate matter, needs a very robust belief in the uniformity of nature. The theory here must be something like this: life existed aeons ago on the hot surface of a cooling planet, the Earth, and that life, as we know it, evolved from single-cell organisms to homo sapiens. Along with matter, some element of mind must have been there at the beginning to account for it evolving into what mind is today.

Let's face it — there had to be a good deal more than an element of mind — there had to be life itself! Bergson's *élan vital* — that force of nature, a force as palpable as gravity, electromagnetism, radioactivity and nuclear, that impulsion for life — must inhabit the universe and make it alive. We are back, it seems, to something like the World-Soul of the Pythagoreans and, indeed, of Plato — not of number or harmony, but an exuberant life-force, something that accounts for consciousness and, perhaps, for a universal animation. Matter and mind are inseparable, but the mind of inanimate matter would be far beyond our comprehension — in the case of a mountain, the duration of a thought being measured, perhaps, in units of a million years rather than being measured in milliseconds in our case. In *The Prelude*, Wordsworth broods on 'unknown modes of being' and on 'huge and mighty forms that do not live like living men', and hails the World-Soul:

> Wisdom and spirit of the universe,
> Thou soul that art the eternity of thought,
> That giv'st to forms and images a breath
> And everlasting motion ...

This is all very fanciful, but it is somewhat along these lines that a theoretical justification of panpsychism must be made. In a more pedestrian way, without going anywhere near so far as claiming life for inanimate matter, we might simply regard mind as an emergent property of matter, in the sense that beyond a certain level of complexity of physical systems new forms appear that require a new level of description. Some elements of the community engrossed in the creation of computer-generated artificial intelligence can see the future appearance of a silicon-transistor mind in electronic neural networks. In physics and chemistry there are numerous examples of new structures emerging as a result of non-linear, dynamic processes — hurricanes in meteorology, solitons in optical communication, electric domains and filaments in conductors. They are relatively transient phenomena and, thankfully, none of them is self-replicating. But, as against the idea of mind being somehow an emergent property in the sense described, mind is most definitely not just the matter of physics, and so could not grow out of something that is. A hurricane is a cyclonic wind that has grown enormously because of a meteorological instability, but it is still a wind. An electric filament has grown into an intense current spark because of an electric instability, but it is still an electric current. Whatever an emergent property of a neural network instability may be, it will not be a mind.

The only sort of mind we know is our own. We impute the existence of something similar in animals, but we cannot believe it to be the same as ours. If we hold to the dual-aspect description, the aspect that is mind must be different in animals from what it is in us, just as the neural structure is different. What the aspect corresponding to mind is for an amoeba is beyond our comprehension, as it is for a mountain, for a forest, or for the whole planet. We are creatures of limited power, and though we see no reason to suppose that amoebae, mountains or planets really have minds, our imagination, using the tools of analogy and extrapolation, see no reason why not. And if we can contemplate minds for amoebae, why not contemplate souls for people? In our various enquiries about the world we inevitably come up against things that are far beyond our mini-human comprehension. Faced with such a situation we can only resort to belief of what is plausible, or to agnosticism, or to indifference. The Principle of Complementarity may ameliorate the mind-body problem, even if it does not solve it. But it says nothing about the soul.

Perhaps the soul is just an icon, an icon symbolising the nature of this human-all-too-human need for comprehension, to understand ourselves and the world we live in, to search for meaning in our lives. In a word, my need, our need, is religious. It is a word scarcely to be taken seriously in today's culture, but the need is deep and it is there and it has always been so and will always be so. Religion takes many forms — the institutional powers of the Church, the Mosque, the Temple; it appears even in secular guise as science, communism, save-the-planet; but there are common strands. I know no better account than that of the mathematician/philosopher Alfred North Whitehead.

> Religion is the vision of something which stands beyond, behind, and within, the passing flux of immediate things; something which is real, and yet waiting to be realised; something which is a remote possibility, and yet the greatest of present facts; something that gives meaning to all that passes, and yet elude apprehension; something whose possession is the final good, and yet is beyond our reach; something which is the ultimate ideal, and the hopeless quest.

Our spiritual life is a manifestation of that hopeless quest.

Chapter Two

A Short History of Animism

I begin my quest with a simple idea of soul, the kind of idea that preceded the ancient philosophic search. People have always had a strong predilection for common-sense, and the common-sense view of soul is that it exists in every one of us. There is simply no doubt that my experience is of an 'I' that is distinct from the body, and it is that 'I' that I call my soul. At its lowest level it is merely a convenient nomenclature that presupposes nothing more that an evident attribute of every human being. A familiar literary example is the use of 'soul' as a synonym of a unit of capital associated with the estate of a Russian aristocrat—so-and-so has an estate of five hundred souls. For soul read serf, but in a Christian country 'soul' sounds better than 'slave' and more redolent of a spiritual sensibility. But the idea of soul carries, and has carried, a good deal more intellectual and supernatural baggage than a simple name-tag. And quite understandably so, given the obvious duality of mind and body.

In common sense usage, soul refers to the essence of the human personality as evidenced by the moral, ethical and intellectual behaviour of the person. And it goes without saying that the moral, ethical and intellectual behaviour of the person can be experienced only when the person is conscious and in full possession of his faculties. It cannot be done when the person is asleep, in a trance or in a coma, yet we must assume that the person still has a soul under those circumstances, or must we? Trances and comas are rare, but sleep is not. What happens to the soul then?

It has been the empirical evidence of dreaming that has literally put life into soul. In dreams we travel, we have adventures, we meet friends and relatives, some of them dead. There are encounters with all sorts of material objects—houses we might recognize, benches, stones. One belief is that during sleep the soul is released from the body to wander about in a world that, at best, is only half-recognizably the one we inhabit. If so it is important not to awake the sleeper too suddenly, lest the wandering soul have no time to return to the body. There is a dream world where dwell shades of people one knows or once knew, and where there are familiar objects—stones, trees, rivers, mountains.

A soul that can detach itself from the body in sleep can surely do the same at the point of death. Belief in ghost-souls brings us firmly into that culture of animism that permeates the history and religions of the world. The rich man dies and his soul goes forth into a new world. That soul needs to be served as well as it was in life, so it follows that his wives and servants must be killed so that their souls can provide a respectable household for the master. And along with them are his favourite dog, his favourite horse, with his favourite drinking cup ready to hand by his dead body. It is a cult of the dead that was practiced widely. Today, ghost-souls hover over ouija boards and queue to invade the medium and have their say in the world of contemporary spiritualism.

The idea of a world that lay beyond our own world that was yet accessible gelled well with the universal belief in the existence of gods. Gods are not to be mocked by modern sceptics—gods explain a lot. How else account for good seasons, bad seasons, storms, drought, fruitfulness and barrenness? And man, primitive or modern—it makes no difference—needs explanations to guide him. Lacking modern science and the belief it has generated, primitive man, where an obvious explanation was elusive, put it down to the action of some god. An unsuccessful hunt might be the result of some offence given to the god of the forest; a successful harvest would require the motherly approval of the goddess of the earth. The gods needed to be worshipped and propitiated with dedicated temples, shrines and libations. The high gods lived in the mysterious world where only the elite warrior could hope to be chosen; other, lower gods, were to be found in this world in groves and forests and lakes. Men once lived in a rich dual world, and some still do. It was a world animated not only by man and living things of all kinds, but also by ghosts, shades, demons and spirits, where nothing could be counted on as being inanimate.

Was there a dual world? What was the essence of the one we could see? Did man have a soul and, if so, was it immortal? What was the connection between mind and body? Questions like these began to be addressed by a few aristocratic men living in Miletus in the Greek colony of Ionia some twenty-six centuries ago, and their thinking changed the civilization of the Western world irrevocably. What was the simplest idea of the essence, the *arche*, of the world? The answer, in spite of common-sense evidence to the contrary, was that, beneath the obvious appearance of things, there existed a single substance that was the fundamental ingredient of all matter. Thus was born the Doctrine of Monism.

What is it about the human mind that entertains ideas like this? The Pre-Socratics may have been the first to articulate the Doctrine, but it is very much alive in modern physics in its search for the Theory of Everything! Is it the need for simplicity, the horror of complexity, the vision of an aesthetic ideal? Does it manifest itself in the belief in One God, or the one true political syetem? Perhaps it is a sign of irretrievable indigenous madness, or of a deep mystical intuition. Whatever is the case, it is an unignorable phenomenon of the human mind, first exhibited by those aristocratic Ionians. In spite of appearances, they argued, the world is but one substance.

How could they possibly be serious! Look around, at the pebbles, at the sand, the rocks. Out there is the sea, with clouds on the horizon and a low sun. Behind are trees and meadows and hills, alive with birds and dogs and squirrels. Feel the breeze and listen to the lapping of the waves, the distant bell. Walk a little way and feel the intense plurality of the world. How could anyone think that in spite of this evident, sumptuous variety, beneath it all there was just one fundamental essence! Madness?

Those articulate Ionians not only believed in the arche of the world, they envisaged it as substance. If there was only one fundamental thing in the world, what was it? For Thales, the first of these great philosophers, the fundamental substance was water, and this was argued for with that rationality that Greek philosophy became famous for. His arguments cannot detain us here since our immediate aim is elsewhere. After him came Anaximander who argued that the fundamental substance could not be something as simple as water but something altogether more mysterious. (Quarks and leptons?) Anaximenes agreed that water was not right but, rather, air was the essential substance. But whatever that divine substance was, it was universal and was therefore as much part of living things as of (apparently) dead ones. Their monism led them to the view that everything was matter and, moreover, that matter was alive. Yet another doctrine, this time the Doctrine of Hylozoism.

But why monism? A glance around reveals countless different things, from stones to trees, from benches to bees, from rain to dogs. Why would anyone believe that each was really, deep down, made of the same stuff? And, moreover, a stuff that embodied life? Yet that was the bold vision of these Milesians — hylozoism, a cosmic theory of matter-life. While the arguments for this or that substance were rational, the intuition that, in spite of appearances, a fundamental substance, be it water, air or something mysterious, existed was wildly irrational. But, in fact, this is the way the human mind works. As the philosopher Bertrand Russell has remarked:

'Even in the most purely logical realm, it is insight that first arrives at what is new.' The Ionian insight gave us monism and hylozoism, and whatever one may think of their doctrines, they have inspired thinkers thereafter.

But what of the soul? Nothing was more wildly irrational in Greece and its colonies at that time than the religion of Orphism and the worship of Dionysus. This was a terrifying and dangerous cult that believed in attracting the intoxicating union with the god—enthusiasm—in sometimes barbaric ceremonies. Knowledge unobtainable in any other way could be had through enthusiasm. The Orphics believed that the soul came from the stars and was part of the World Soul. As such it was transferable between living things and the transmigration of souls meant that bad souls, after death, could be reincarnated in some lower life-form. It was a mystical cult that convinced and excited Pythagoras and his followers in the Greek colony of Croton in Southern Italy. The Pythagoreans had a remarkable intuition of a different sort of monism of the world, namely that it was Number. Pythagoras has the reputation of discovering his famous theorem of the right-angle triangle and has become famous as the first mathematical physicist, seeing Number as the essential thing to understand the universe. In fact, Pythagoras was much more of a mystic than a proper mathematician, but his reputation in modern science remains intact. For example, the number 1 represented the Pythagorean Unity of the World and the first triangular number; 2 represented diversity and disagreement; 3 was the second triangular number; 4 justice (square deal); 5, being the sum of the first even and odd numbers, represented woman and marriage (1 is too holy to be odd!); 6 the second triangular number and the base of the Beast (666); 7 the number of heavens on one count; 8 the cube; 9 the number of heavens on another count; 10, the sum of the first four numbers represented perfection, and being the fourth triangular number, represented by the revered sacred symbol, the tetractys, the triangular figure composed one, two, three and four dots.

Another aristocratic Ionian, this time from Ephesus, was Heraclitus, a somewhat bombastic mystic. His ideas would certainly not appeal to the tender-minded of today with his low view of humanity. Only force, he thought, would compel them to act for their own good, an opinion shared some twenty centuries later by the philosopher Thomas Hobbes. Not a democrat, then, more an autocrat. Not surprisingly, his fundamental substance was fire. All was change. Everything was in a permanent flux—there was nothing eternal. 'You cannot step into the same river twice', goes his famous remark. He was, like his Milesian fellow-countrymen, a robust monist, so much so that he regarded opposites as the same thing! With strife opposites mingle to create harmony. Good and evil are one; the way up and the way down is one and the same. Soul is a mixture of fire and water, the fire being noble, water ignoble. Asceticism keeps the soul dry, which is good since it is then full of fire. Only noble souls can hope to survive death. Above all, Heraclitus is most famous for his insight that the essential feature of the world is Becoming rather than Being.

In the south of Italy, Parmenides, living in Elea, couldn't believe there could be such Ionian nonsense. With impeccable logic Parmenides had convinced himself that change was utterly illusory. What truly *is*, is timeless. Later, Aristotle summed this up crisply: 'What *is* does not come into being, for it *is* already; and nothing could come into being from what is not.' Thus Being and not Becoming is the truth. Furthermore, degrees of being, implied by the Milesians, is impossible since what exists is indivisible and continuous. There can be no void. There are only two things: what is, and what is not. The appearance of change is some sort of hallucination. This doctrine, reinforced by other Eleatic philosophers like Zeno and Melissus, had a strong influence on Plato. Zeno, for example, famously advanced several arguments for the impossibility of motion, which has impressed everyone ever since.

(The dispute between Heraclitus and the Eleatics presaged the modern conflict between Henri Bergson and physics, the one extolling creative Becoming, the other describing nothing but Being. A case of *plus ça change, plus c'est la meme chose*—in more ways than one.)

Pythagorean animism differed from the Milesian version through its emphasis on the human soul. Encountering this doctrine, with its addition of number to the various fundamental essences of the Ionians, the Sicilian, Empedocles proposed a radically new approach. First of all, this strange idea of the Oneness of the world must be abandoned: monism must go. Nevertheless, the idea of fundamental substances was good. Empedocles proposed that the world and all that was in it was composed of four elements—earth, water, air and fire—and his proposal was accepted by thinkers up to the modern era. The conflict between the One and the Many was resolved by seeing Being as One *and* Many. But monism has been too powerful an attraction to ever be given up and the conflict between monism and pluralism rages still. Empedocles was extraordinary in a number of ways. From the study of eclipses he knew that the earth was round. Perhaps the most striking of his ideas came from his studies of plants and animals—he proposed a theory of evolution, including the survival of the fittest. Unhappily, his idea was given short shrift by Aristotle and it never got off the ground. What bothered Aristotle was the fantastic set of monsters that Empedocles introduced in support of his theory. He imagined heads without necks, eyes without foreheads, solitary limbs without a body, all of them striving to find what they lacked and become complete—those that failed were doomed to extinction, those that succeeded survived. We had to wait for Darwin.

Evidence that monism was far from dead came once again from the Ionian tradition. Anaxagoras proposed that mind (nous) was the fundamental substance. This was a very different idea from what had gone before since mind was quite distinct from matter. The materialism that was at the core of the Ionian tradition was abandoned. Anaxagoras maintained that it was mind that distinguished living from non-living things, but crucially only in proportion that it inheres in the thing. Mind was the source of motion, but he taught that in everything there was a portion of all that exists, including mind, and that all that exists is infinitely divisible. Thus apparently inanimate objects all contain some mind, be it infinitesimal. In this, Anaxagoras diverges from the hylozoism of the earlier Ionians. As Aristotle and other philosophers remarked, after getting the universe going with mind, Anaxagoras did not use 'mind' to describe the motion of ordinary objects. Nevertheless, Anaxagoras, however tentatively, may be said to have initiated the Doctrine of Panpsychism — the teaching that every concrete thing has a mind in some sense. Once again, this was a teaching that did not die; it was alive and kicking in the nineteenth century in the writings of that eccentric physicist Gustav Fechner, as we will see.

The cascade of ideas about the nature of the world that flowed from the Ionians, the Pythagoreans and the Eleatics was unprecedented. Doctrines born of one group triggered opposing doctrines in another. The monism of the Ionians, the Pythagoreans and Heraclitus suffered severe criticism at the hands of Parmenides, Empedocles and Anaxagoras. The conflict between Being and Becoming raged between Parmenides and Heraclitus. Through it all, the soul, the essence of living things, survived as a kind of mysterious, immaterial being. Now, inevitably, there are minds that are troubled by mysterious, immaterial things and strange abstractions and prefer to embrace a more down-to-earth view of the world.

The time being what it was and the place being Greece, it was also inevitable that the more down-to-earth view that emerged was utterly unique and staggeringly prescient. This was the solidly materialist world picture of Leucippus (Milesian?) and his pupil Democritus from Abdera in Thrace. Everything was matter and substances were not infinitely divisible as Parmenides and Anaxagoras claimed; matter was composed of atoms forever in motion in the void. There were an infinite number and kinds of atom moving according to deterministic laws. The mind, the soul, is composed of especially fine atoms that dissipate after death, so no survival after death. Moreover, since atomic motion is absolutely determined there can be no free will. Epicurus, much later, took up this atomic theory, but not liking the absolute determinism of Democritian theory allowed the atoms occasionally to move of their own volition, from which free will would be saved. A delightful account of atomic theory is to be found in the book by the Roman, Lucretius, entitled 'De Rerum Natura' (though Cicero labelled it more as poetry than true philosophy). From Leucippus and Democritus we have the first Doctrine of Materialism.

The contribution to the culture of Western civilization made by the Pre-Socratics is breath-taking. The innovative application of reason to the most intractable mysteries of experience would be enough in itself for unbounded admiration, but the breadth and ingenuity of the flow of new ideas make their contribution prodigious, enough to excite the minds of untold generations of thinkers. Among the first of these was Plato, the pupil of Socrates in Athens.

The ethos of Athens at the time was not remarkably conducive to the teaching of philosophy, having forced Socrates into suicide, but in spite of that Plato founded his Academy there. The materialism of Democritus was definitely not for him. Such was his mysticism and teachings concerning the soul, that much of his thought was taken over into Christianity hundreds of years later. Plato saw the essential metaphysical problems presented to man by the world in terms of dualisms—reality versus appearance, ideas versus the evidence of the senses, the souls versus the body. Following Parmenides, he saw reality as something unchanging that underlay appearance, an eternal Being rather than a meaningless flux of Becoming. For him, like Pythagoras, number and mathematical truths existed as eternal Forms, independent of mind and matter. The fact that a right-angled triangle possessed sides whose lengths obeyed the theorem of Pythagoras must have existed long before the theorem was discovered. Today, more than two thousand years on, mathematicians are divided into Platonists who believe in the eternal existence of mathematical truths and Non-Platonists who believe that a mathematical truth comes into being only when it is demonstrated by a mathematician. Plato believed that everything had its eternal element, its Form, which transcended the individual thing. There are beautiful things and there is the idea of beauty. One had to distinguish between the man who loved beautiful things and the man who loves the Idea of Beauty. The former has only opinion, the latter has knowledge.

Plato saw the created universe as an exhaustive replica of the World of Ideas. Since this universe was the only one created, all possibilities had to have been realised. Everything with a genuine potential for existence, animate and inanimate, Forms of every kind, and of course mathematical truths, all had to be a part of this world. How many kinds of being must this world contain? The answer Plato gives us is — all possible kinds! The historian Arthur Lovejoy has called this Plato's Principle of Plenitude.

Plato's influence today, outside of mathematics and his Theory of Forms, is still felt strongly in the world of politics. His account of a utopian society described in the 'Republic' is famous, or infamous, depending on your point of view, for his advocacy of an ideal government by elite, all-wise philosophers. As in communism, the state is prior to the individual and there is little private property, and that includes wives, and children looked after by the state. There is more than a touch of the influence of Sparta here. The 'Republic' was written with the aim, not primarily to advocate an utopian polity, but to define justice. Plato was especially concerned to oppose the definition offered by Thrasymachus (a character in the dialogue but also a real person). Thrasymachus had a view of justice that is manifested by every tyrant and dictator down to the present day, namely, that justice was the interest of the strongest. For Plato, it was everybody getting on with their own work and not interfering with other people. Not much of that around today! But also, he did not believe that a society could be strong without the belief in a living personal God. Like the philosopher Friedrich Nietzsche, who vainly attempted to reach his contemporaries two thousand years later through the question: where is the source of morality when God is dead?

Plato believed that all things with opposites are generated from their opposite, hence there is life after death. All things that are complex dissolve, the soul being indivisible survives. The soul of the true philosopher departs to the invisible world of the gods. The impure soul, a product of the fleshy lusts of the body, will become a ghost or will enter the body of an animal, the particular type of animal being dictated by its character. Here, primitive animism has been overlain with the Pythagorean idea of the transmigration of souls and taken well on the way to religion. Suitably modified, it was, indeed, taken over into Christianity.

The Orphic mysticism that Plato shared with Pythagoras became heavily diluted in its transmission to Aristotle, the most brilliant student of the Academy. Hailing from the kingdom of Macedonia in Stagira (a part of the world the Athenians would regard as barbaric) Aristotle's fame as the star of the Academy spread to his home territory and earned him the privilege of becoming tutor to a wild youth who was to become world famous as Alexander the Great. Afterwards he returned to a not-very-welcome Athens to establish a school which was called the Lyceum. The name derived from its situation in the grounds of the temple to Athene-Lyceus (protector against the wolf-lycos). In the grounds of the temple there was a Walk known as Peripatos along which Aristotle and his pupils used to stroll along in deep discussions, and so the school became known as the Peripatetic School and its pupils as Peripatetics.

The philosophy of Aristotle is sometimes described as common sense Plato. Aristotle took on Plato's idea of Form but made the Form to be the specific essential attribute of the thing, not a mystical universal. For example, there is redness only because there are red things. The state was not some utopian entity, but an organism designed for the good life and the possibility of noble acts. The best governments were monarchy, aristocracy and constitutional polity; and certainly not oligarchy, tyranny or democracy. His delightful idea of the state is that it should be no larger than the land surveyed from a hill-top, so that all citizens would know one another. All we can say now is, dream on!

The idea of substance crops up again. Aristotle identified three kinds of substance: sensible and perishable (plants and animals); sensible but not perishable (the heavenly bodies, each one of them alive); neither sensible nor perishable (the souls and God). But the soul is no longer simple. As part of the body, it has a vegetative and appetitive side which is deeply irrational. This part dies with the body. The higher part of the soul is mind, the part that understands timeless things being rational and contemplative. This part is divine and does not die. This implies that what survives is utterly impersonal. His God is also impersonal. Aristotle wondered about the fact of motion. He saw that one motion was caused by some interaction, some other motion, but how did motion originate in the first place? He therefore introduced God as the First Mover. Not exactly your personal God!

Aristotle was deeply impressed by the infinite variety of life forms, and he believed there was a cosmic order that connected the higher forms to the lower forms. The sequence had minute gradations, so minute that it was impossible to say where the transition took place between living and non-living. He could not, therefore, rule out animism at all levels. With steps in the sequence so small that they were virtually infinitesimal, Aristotle's cosmic order was a continuous sequence that harmonized with Plato's Principle of Plenitude—all possible life-forms existed. The order stretched from the apparently inanimate, through the lower forms of life, up through the animals to man, and beyond to the angels and the aetherial beings. This grand vision was recognized by generations that came after Aristotle as the Great Chain of Being. Its development reached its peak in the Neoplatonic fever of the renaissance.

Let us sum up the contributions that the Presocratics and their philosophical heirs made to the debate about body and soul, especially as the debate rumbles on today. Is the body purely physical, that is to say, material, or is it spiritual, that is to say, immaterial? Is the soul purely physical, or is it spiritual? The Greeks generally regarded the body as matter, but they needed the soul to account for the power to instigate motion and for perception and thinking. But what was its nature? Imprisoned god? A fragment of the arche of the world? The Milesians seemed not to distinguish a category of spirit different from matter, so, for them, all was matter, with the soul perhaps being a rarified form of the fundamental substance that made up everything. For the atomists the soul-atoms were spherical. Since the like can know only the like, Empedocles had the soul made up of all the four elements so that water, fire, air and earth could be perceived. The Pythagoreans, on the other hand, having more of a mystical bent, emphatically took the soul to be spiritual and distinct from matter. So did Plato and, with caveats, so did Aristotle. One of the caveats was that soul had other, more down-to-earth, attributes, one of which was that it referred to the function of the body. Only spiritual souls survive death, physical and functional souls do not. Looking forward to the modern debate, we can identify six ways in which body and soul have been regarded. They are summarized in the table below.

BODY	SOUL	DOCTRINE
Physical	Physical	Presocratic
Spiritual	Spiritual	Idealism
Physical	Spiritual	Dualism
Spiritual	Physical	
Physical	Non-existent	Materialism
Physical	Function	Functionalism

We await the new philosophy that argues for a spiritual body and a material soul.

In all this, it is striking that the Greek thinkers had no doubts about the existence of the soul; the questions were: what was its nature? Was it material? If so, what was it made of? Was it spiritual? If so, did it survive death? Did the World have a soul? Did the soul have parts? A broad spectrum of opinions was argued for, ranging from those of the relatively unsophisticated Presocratics to the prescient rational analyses of Plato and Aristotle, analyses that have provided the foundation of much of our modern ideas. And threading through it all are the mysteries of being and becoming, that bear upon the nature of the soul. Not only a broad spectrum, but a rich one. A belief in the existence of soul, the belief in a generally animated world, made for a spectacularly stimulating life of the mind that reached its zenith in the Neoplatonic doctrines and the Hermetic ferment of the renaissance.

Chapter Three

A Magical World

The Milesians thought that, in spite of all outward appearances, there was only one world substance. Though they could not agree about the nature of that world substance, they did agree that it was eternal and divine and therefore alive. But what could possibly have convinced them that apparently inanimate objects were, in fact, alive? Was it some quasi-religious or mystical feeling connected with the idea of a World Soul in which everything in the world had a place? Or was it the logical consequence of their monism, the idea of a universal substance that was part of living things and therefore, in some sense, living? Rationally it was more likely to be the latter rather than the former, though a mystical strand cannot be ruled out. But in addition to any logic, there was, in fact, hard empirical evidence! It was generally known that there existed lumps of matter that could cause other lumps of matter to move, so surely those lumps of matter were alive. Making things move was a property that only living things had. As elements of the divine world stuff they were proof that world stuff was alive. Those lumps of matter that could exert invisible magical forces were the lodestone and amber. The lodestone is the old name for magnetite, an ore from a region in present-day Turkey known as Magnesi; amber is solidified resin of estuarian origin whose Greek name is electron.

Motion is produced by one thing bumping into another, or being pushed or shoved, or being thrown. There is always contact between one body and the other. How could it be otherwise? Amber and the lodestone showed that it could be otherwise. Motion could be generated without things touching. Attraction and repulsion could operate and induce motion of things that were not in contact with the amber or lodestone. How was this possible? When gravity was discovered by Isaac Newton, it also had the magical property of attracting things at a distance. To Descartes, this was unacceptable — somehow the planets had to be pushed around in their orbits. Action at a distance was supernatural. To the ancients, the action of amber and the lodestone was indeed magical.

And their properties are indeed strange: invisible forces that stretch across space. Just play with kits available in every toy shop, the ones that have tiny strong magnets, and the effect is palpable. Unless one is blinded by the science of electromagnetism to the extent that its explanation of 'how' has driven out the wonder of 'why', these invisible actions on other matter seem as magical now as it ever did. And how much more magical would it have appeared then! The lodestone attracts iron and if you rub amber it attracts all sorts of chaff, and, what's more, both attractions function at a distance. In some circumstances attraction is replaced by repulsion, for example, rubbed glass can repel amber and, today, we understand how this works in terms of magnetic and electric fields, both of which are related to the properties of a minute particle we have christened the electron. Nobody knows what an electron is, only what it does. In Milesian terms the electron would be an element of the world stuff and would lose nothing of its divinity thereby. Let us contemplate for the moment the beautiful idea of the divine electron. An adequately reformed science will surely take on board the divinity of leptons and quarks.

The action of those invisible forces must have had a profound effect on ancient philosophic thought. And not only on the Pre-Socratic mind; Aristotle himself believed that desire directly attracts inanimate objects, causing the objects to move towards their proper places in the cosmic scheme of things, rather as a living creature grows or moves towards what it desires. I think it is striking that among thinkers that were famous for their rationality, animism, in however limited a form, was accepted as established. And not only on the earth but also in the heavens. For both Plato and Aristotle the stars were superhuman intelligences, with subordinate gods incorporated in the spheres of the seven heavenly bodies, Moon, Venus, Mercury, Sun, Mars, Jupiter and Saturn. With this belief, incidentally, astrology makes a lot of sense; these gods were bound to influence activities on Earth. Alchemy, the messy chemical activity that was mystically motivated by the search for the elixir of life and the so-called philosopher's stone, the stone that could convert lead to gold, regularly called on those ethereal gods for guidance. Astrology and the study of the heavens was most advanced in Babylonia, and after the conquest of Alexander, much of this knowledge must have become more widely spread in the Hellenic world.

That same remarkable conquest must also have resulted in the importation the heady strands of oriental religion and mysticism. The Jewish religion with its single God must have appeared incomprehensible to the Greco-Roman occupiers of the Middle East. The acceptance by those conquerors of the concept of a single all-powerful God had to await the birth of Christianity. Only in the early Middle Ages, when the cabala, that mystical Jewish cosmology, partly Neo-Pythagorean in flavour, became known, was there a significant input of Jewish thought. Arts of divination, practised by the magi, the wise men of Babylon and Persia, entered the culture of the West and initiated a movement that was to trouble both priests and practitioners in the Christian era to come. Christianity was spread at first by Jews preaching to Jews and, thereafter, Jews preaching to gentiles. In the first few centuries of the Christian era the intellectual climate of the Greco-Roman world was wildly unsettled, first by Christianity itself, but also by a storm of Greek animism, astrology, divination and, much later, the cabala. In Christianity, heresy followed heresy, now seen as such after the true religion was defined by committee in the fourth century. In philosophy, as a last example of Greek rationality, the writings of Plotinus appeared, and was followed by the beginnings of that complex and mystical doctrine that became known as Neo-Platonism.

In Neo-Platonism, a syncretism of Plato, Aristotle and Christianity expounded by Plotinus in the third century AD, there are two new ideas, emanation and empathy. In the Aristotelian hierarchy the essence of a higher level emanates downwards, and so the whole world is filled with the divine emanation from God. Such an idea has disappeared from our present-day culture, leaving, somewhat bathetically, only a materialistic trace perhaps in the cosmic microwave background. The idea of cosmic empathy, on the other hand, still survives in the belief that somehow we have an insight into and an understanding of the workings of nature, and that what science describes really exists in some sense. Its expression in Neoplatonism is the assertion that each being in the world contains all within itself and at the same time sees all in every other, so that everywhere there is all; the microcosm contains the macrocosm. Like a hologram, each little bit contains information about the whole.

Plotinus was born in Egypt, studied in Alexandria and finally settled in Rome. While admiring Aristotle, he regarded Plato as by far the greater, and he expounded Plato's ideas with a clarity not always to be found in the Master's dialogues. The soul, for Plotinus, was Essence and being a Platonic Idea was eternal and therefore immortal. The pure world of Essence, or the All-Soul, contained all souls that existed. In that world, no soul was separated from other souls; contact with other souls would be lost only when a soul was joined to a body. Following the Platonic and Aristotelian party line, Plotinus advocated contemplation over action as a way of life most conducive to re-contacting the Essence. We will come across some idea of what was meant by 'contemplation' later. His student, Porphyry, published the writings of Plotinus and it is that doctrine, heavily laced with mysticism of Porphyry's own writings, that has come down to us as Neo-Platonism. So what was, or rather, what is Neo-Platonism?

The soul was defined by Socrates and endorsed by Plato. Socrates taught that the soul was the moral, intellectual and responsible agent of man, quite separate from the body, more a spiritual entity than the material, albeit rarified, soul of the Milesians. Plotinus believed in this soul, but, unlike the Christians, he would have nothing to do with, what to him and his Greek mind, was the utterly incomprehensible notion of sin and redemption. There was the quest for Virtue, the quest for the Good, but no Jewish sin. His universe consisted of three Ideas: the One, Intellect and Soul. These were Platonic ideas in spiritual form. The One is God and beyond knowing (to be endorsed emphatically centuries later by the philosopher Emanuel Kant). Intellect is the Divine Mind containing all Ideas, including all Intellectual Forms that exist in the lower spheres. Soul is Universal Soul, the All-Soul, which is an emanation from Intellect. As Intellect contemplates the One and emanates its influence downwards, so Soul contemplates Intellect and emanates its influence downwards, generating the Natural Universe. Each man's soul is part of the All-Soul and the All-Soul is in each soul in three ways. First is the Intellectual Soul, utterly untouched by matter and accessible to man only if he becomes god-like through appropriate contemplation. Second is the Reasoning Soul, the principle that constitutes the normal man, separable from the body but not separated. The last is the Unreasoning Soul, the animating principle of vegetative, nutritive and generative faculties. Plotinus's vast vision of a Spiritual Universe, his spiritual hierarchy, the emanation of forces from higher to lower, was essentially a syncretism of late Greek philosophy. It was to have a profound influence on Renaissance Man when it was rediscovered after it had disappeared into the libraries of monasteries and of Islam during the barbaric Dark Ages.

But was it not yet another effusion of madness of the human mind! That need for an all-embracing cosmic structure stimulated an adrenal-like fuelling of flight from a meaningless universe, and it produced a powerful stimulant of the imagination that was to transform the intellectual environment of the renaissance, and beyond. Mad or not, it was to become the midwife of science itself.

At the time, Neo-Platonism had a significant influence on the development of Christianity. The concept of an immaterial soul was eagerly adopted by the Church, but it had to be a soul that was sinful. As such it needed the care of the priesthood through whom, and only through whom, it could be redeemed and saved from an eternity in Hell. The business of Superhuman Intelligences that were the stars and planets was tackled by converting them to Beings divorced from cosmic matter. The pantheon of Plotinus was replaced by the nine orders of Spiritual Beings; Cherubim, Seraphim, Thrones, Dominions, Powers, Virtues, Principalities, Archangels, Angels.

But those pagan visions were not to be extinguished. Man's soul in some may well be the pious, God-fearing darling of the priest, but in others it was felt to be god-like itself. Contemplation upwards gave man the power to experience the ecstasy and enthusiasm of experiencing God directly. In those early centuries of Christianity, religion and NeoPlatonism existed side by side, and not uneasily. There would come a time when such cohabitation would be anathema to a powerful Church, but these were early days. The doctrines of Plato and Aristotle held great sway. They contained views of the soul and of the cosmos that were not absolutely inimical to the Christians of the time. Plato's elevation of the soul over the body was particularly attractive but, equally, the Christian concepts of sin and also Incarnation had to be deeply unattractive to any follower of Plato. One merely had to savour the intoxicating flavour of some Neo-Platonic writers in order to predict with great confidence (especially with hindsight) that any accommodation of Christian belief and Neo-Platonism, if possible at all, was bound to be short-lived.

Here is Macrobius, high administrator to the emperor Honorius, writing in the fifth century:

> Since from the Supreme God Mind arises, and from Mind, Soul, and since this in turn creates all subsequent things and fills them all with life, and since this single radiance illumines all and is reflected in each, as a single face might be reflected in many mirrors placed in series; and since all things follow in continuous succession, degenerating in sequence to the very bottom of the series, the attentive observer will discover a connection of parts, from the Supreme God down to the last dregs of things, mutually linked together without a break. And this is Homer's golden chain, which God, he says, bade hang down from heaven to earth.

In the first couple of lines he has encapsulated Plotinus's account of Plato, and in the last few lines he has done the same for his account of Aristotle. For Macrobius there is equal music regarding Plato and Aristotle, but for those early Christian thinkers like St Augustine and Boethius, Plato was where it was at, whereas Aristotle (good man, bless him) was not in the same class. It took until the late Middle Ages for Christian scholastics like St Thomas Aquinas to extol Aristotle. By which time it was a bad idea, as far as physics was concerned, but that is another story.

But Neo-Platonism certainly could stimulate the literary imagination, notably Dante's Paradiso. In his *Convivio*, a banquet of ideas, Dante goes further and manages to relate the seven planetary beings with the medieval curriculum. The Moon, he says, is grammar, Mercury, dialectic and Venus, rhetoric. These make up the Trivium. The Quadrivium is arithmetic (Sun), music (Mars), geometry (Jupiter) and astronomy (Saturn). His arguments for these assignments are best left unaired out of charity.

As though the bubbling intellectual activities of the early Christian era weren't enough, documents appeared that purported to reveal the ancient doctrines of the Egyptian god, Thoth, the scribe of the gods and the god of wisdom. Thoth was identified with the Greek god Hermes and his vast output earned him the name of Hermes Trismagistus (Thrice Great). This literature, the Corpus Hermeticum, described the arts of astrology and of the occult sciences, the secret powers of plants and stones, and the magic that was made possible in a world where everything was alive. This knowledge, attributed to Hermes, was believed to be very ancient and it was therefore deeply revered. For over twelve centuries it was accepted at its face value and it became yet another profound influence on fifteenth-century Renaissance thinkers like Marsilio Ficino and Giovanni Pico della Mirandola. Its trenchant revival of a full-blooded animism and the magical practices consequent on the belief in panpsychism must have made the Hermetica an uncomfortable source of wisdom running parallel with the teachings of the Christian Fathers. Accounts that it contained of speaking statues, airy spirits, demons spiritually between men and gods, couldn't have gone down well with the Church, especially as a lot of it was clear kin to the worship of saints and angels. Moreover, the Hermetic definition of God: *Sphaera infinita cuius centrum est ubique, circumferential nusquam* ('Infinite sphere of which the centre is everywhere, and the circumference nowhere'), was far too *abstract*.

A vibrant magical world was portrayed that makes our materialistic world seem very boring in comparison. It had the following attributes;

1. The world is a unity. Its apparent plurality is an illusion. (Monism never dies!)
2. There is a cosmic hierarchy, the hierarchy of the spheres.

3. Emanation — ideas, forces, spirit at a higher level finds reification at the lower level.
4. Correspondence — as above, so below.
5. Macrocosm-microcosm — man has something of everything within him.
6. The Principle of Plenitude — all that is possible must exist.
7. Theory of knowledge — the like can be known only by the like.
8. The teleological world — all change is purposeful.
9. Animism — everything is imbued with spirit.
10. Sympathies — things separated in space and time can be connected. (Action at a distance!)

Much of this is plainly Neo-Platonic and there were other seams in the Hermetica that had a Christian flavour. We learn from those marvellous accounts of renaissance magic by Frances Yates that, eventually in 1614, Isaac Casaubon concluded that the Hermetica could not possibly be as ancient as claimed but was written by various, probably Christian, authors, sometime in the early centuries of the Christian era. But by Casaubon's time the authority of Hermes Trismagistus had been accepted long ago when some of the Corpus Hermetica turned up in Florence around 1463. An excited Cosima da Medici ordered Ficino to drop everything and produce a translation from the Greek. What he found fascinated him and other Renaissance thinkers. A flavour of some of the wild, somewhat blasphemous hubris that it contained can be had from the account of how Hermes Trismagistus saw the mind.

> See what power, what swiftness you posses. It is so that you must conceive of God; all that is, He contains within himself like thoughts, world, Himself, the All. Therefore unless you make yourself equal to God, you cannot understand God: for the like is not intelligible save to the like. Make yourself grow to a greatness beyond measure, by a bound free yourself

from the body; raise yourself above all time, become Eternity; then you understand God. Believe that nothing is impossible for you, think yourself immortal and capable of understanding all, all arts, all sciences, the nature of every living being. Mount higher than the highest height; descend lower than the lowest depth. Draw into yourself all sensations of everything created, fire and water, dry and moist, imagining that you are everywhere, on earth, in the sea, in the sky, that you are not yet born, in the maternal womb, adolescent, old, dead, beyond death. If you embrace in your thought all things at once, times, places, substances, qualities, quantities, you may understand God.

This is more a description of the attributes of God rather than a realisable programme for getting to know God. But it points the mind in god-like directions, it fires the imagination. It recalls an episode in one of Richard Jeffries novels (Bevis, maybe) where the protagonist sprawls on a grassy hill and sends out his imagination to the distant sea, imagining himself there in detail. Marsilio Ficino, physician and priest, in a letter, exhorts the soul to free itself of the body.

Know thyself, O divine race, clothed in mortal raiment; strip thyself, I beseech thee, in so far as thou canst, nay more; I say, with thine utmost endeavour separate the soul from the body, and reason from the affections of the senses. Then straightway thou shalt see the pure gold freed from the defilements of earth, thou shalt see clean air when the clouds are dispersed; then, believe me, thou shalt reflect thyself as sempiternal ray of the divine sun ... Yet thou believest thyself to be in the lowest part of the world, because thou seest not thyself soaring above the heavens, but only thy shadow, the body, in a lowly place. It is as if ... a bird, flying through the air, should think itself on earth while it sees its shadow flying on the ground. Therefore, forsaking these

> shadows, return unto thyself, for this shalt thou return to greatness.

Perhaps this was what the Greeks meant when they advocated 'contemplation' as the path to the Good. A younger contemporary of Ficino, Giovanni Pico della Mirandola was distinctly more bombastic than his erstwhile teacher and had no qualms about mixing the cabala with the gentler magic of Ficino. His *Dignity of Man* (1487), a planned oration, pulls very few punches.

> Let a certain holy ambition invade our souls, so that, not content with the mediocre, we shall pant after the highest, and (since we may if we wish) toil with all our strength to follow it. Let us disdain earthly things, strive for heavenly things, and finally, esteeming less whatever is in the world, hasten to that court which is beyond the world and nearest to the Godhead. There, as the sacred mysteries relate, Seraphim, Cherubim and Thrones hold the first place; let us, incapable of yielding to them and intolerant of a lower place, emulate both their dignity and their glory. Since we have willed it we shall be second to them in nothing.
>
> How must we proceed ...

Indeed, he may well ask! And in his *Conclusiones*:

> Nothing in the world is devoid of life ... Wherever there is life, there is soul; wherever there is a soul, there is a mind ... Nothing in the universe is subject to death or corruption. Wherever there is life, there is Providence and immortality.

After Pico, the cabala and the Hermetica became fused into a single heady brew.

The contrast with the teachings of the Church with its emphasis on sin and salvation could hardly be starker. Here was a fanfare of a thousand trumpets, trombones and timpani for man to cast off a soporific humility and glorify in his god-like essence. It was call to go out and master all the occult forces, to grasp a deep understanding of the world, to manipulate all spiritual powers, in a word to be a god. If the teachings of Christianity mirror Nietzsche's slave morality, the teachings of Hermes Trismagistus utterly outshine his master morality.

But, with Nietzsche, who can deny that the Hermetic fervour was *the* driving force for science? Nor did it die once science began. William Gilbert of Colchester (1540–1603), physician to the Royal court, and the first to apply science to the study of the lodestone and to amber, says in his famous book, *De Magnete* (1600):

> We consider that the whole universe is animated, and that all the globes, all the stars, and also the noble earth have been governed since the beginning by their own appointed souls and have the motives of self-conservation.

It was an exciting world. It was a world full of hidden powers as exemplified by amber and the lodestone. Talismans could be made in ways that drew down the planetary powers, the pale imitations of which swing as mascots in the back windows of Fords and Hondas today. The stars determined your life, and still do in today's tabloid press. Wear gold to attract the vitality of the Sun. Wear dull colours for study to attract the asceticism of Saturn. Wear purple to attract the gravitas of Jupiter, etc. It was a supernatural world, an irrational world, a religious world.

And animism was by no means dead. Among those minerals that were being classified there were 'stones' with occult properties. For reasons quite apart from their occult properties, crystals have always exerted a fascination. In a world where by far the majority of naturally occurring solid objects have shapes that are quite random — pebbles on a beach, chunks of rock, bits of wood — crystals with their regular shapes were bound to stand out. And coveted as gemstones. Anyone working with them could not be other than struck by the precise angles the facets made with each other and the regularity of form.

It must surely have been the case with the young Pythagoras (c. 580–500 BC) whose father was a gem-engraver and who would have followed in his father's trade according to Greek custom. Scarcely a coincidence, therefore, that he was the first to distinguish the existence of all five regular solids — pyramid (four sides), cube (six sides), octahedron (eight sides), dodecahedron (twelve sides) and icosahedron (twenty sides). Cubic crystals are common — rock salt, fluospar, diamond. Spinel exhibits well-defined octahedra and one form of iron pyrites is dodecahedral (found, as it happens, in South Italy where Pythagoras lived). Pyramids tend to be associated with a cube or capping a square column, or twinned in the form of an octahedron. If a crystal with icosahedral form exists it must be rare. The fact that Pythagoras saw that such a regular solid was a logical possibility points to his purely mathematical prowess.

Apart from the desirability of gemstones as ornaments there was the belief in some minds that their regularity of form in a world of random shapes signified some occult meaning. Back to magic again! Given that the use of the term 'stone' for gems and crystals alike was commonplace, the search by the alchemist for the 'philosopher's stone' that could convert dross into gold and keep him perpetually young was really a search for the ultimate, magic crystal. (One might idly speculate that the philosopher's stone is an icosahedral crystal, perhaps!) Lesser stones still enchanted. Their internal radiance suggested the presence of magical powers that could act as charms against evil or as cures for ailments. Adepts, gazing into the crystal in the activity known as scrying, claimed to see images that told of the past or that predicted the future (though there is no evidence that the striking technological future of crystals as transistors, lasers etc., was ever foretold). Such was the fascination with gems and their magic that whole books were written about 'stones' — lapidaries — and they continued to be written well into the seventeenth century.

And it was all going to change. As Science and the Enlightenment got under way, the world began to lose a sense of freedom and to look increasingly purposeless. Some myths were strong enough to retain some power, astrology for one, and some retained the power to inspire, for example, Gustav Holst's *Planet Suit*, but for most, they either disappeared or were transformed: scrying became crystallography, alchemy became chemistry. Many of the first scientists in the seventeenth century, including Isaac Newton (1642–1727), took alchemy seriously and pursued their science within a recognizable aura of mysticism. But a sterner, cooler spirit, if still devoutly religious, became the norm, and the world began to lose its soul. In order to appreciate the magnitude of the change, we need in the next three chapters to recapitulate the history of the New Science, its exploitation in technology, and its strange evolution towards a kind of mathematical theology.

Chapter Four

The New Science

It was in the sixteenth century that the intellectual life of the world began to change in a deep and irreversible way. Beginning in Europe and largely confined there until recent times, it was a change in the way we perceive the world and our intrinsic humanity that has affected every thinking person on the planet today. Its practical consequences have affected everyone, thinking person or not. Europe seemed to suddenly realise there were things to be done with all that arcane knowledge of the ancient world. It just needed a new perspective, a realisation that the world and its nature might be understood in ways other than via ancient texts. Received knowledge from the ancients may indeed be worthy, but it was not the only knowledge possible. The world might be understood through processes that mixed intuition and that messy trial-and-error activity more familiar to the craftsman of one sort and another than the scholar. They were processes that became more and more sophisticated, crystallizing over time into what we now call the scientific method.

It is difficult to exaggerate the trauma suffered by the intellectual religious life of the time by these events. Scholars pored over the works of Aristotle; theologians pondered the Christian philosophies of St Augustine and St Thomas Aquinas; more raffish philosophers considered the emanations that proceeded from the heavenly Empyrium to lowly earth.

The nature of the world was to be found in the literature of God and the ancients. Understanding of God's creation was to be had through textual perusal and rhetoric. The last thing, if ever, a member of the academic and theological elite contemplated was to contemplate the possibility of change, of evolution, much less to engage in the grubby activity of experiment in the study of change. (It may be said that little is different today. Theoreticians rarely get their hands dirty in the laboratory.)

Yet the Hermetic magus still existed throughout the birth of science, his influence in literature evident in the play by Christopher Marlow about that famous example, Dr John Faustus, and also as Prospero in Shakespeare's *Tempest*. Around 1600 the mathematician, Dr John Dee, and his spirit medium, Edward Kelly, toured the courts of Europe, demonstrating the existence of the spirit world, and there was Robert Fludd, magus extraordinary, increasing the Hermetic cosmology to twenty-two spheres. A secret society, the Rosicrucians, Brothers of the Rosy Cross, whose members probably included Fludd and may even have included the Lord Chancellor of England, Sir Francis Bacon, publically claimed god-like wisdom and power through arcane knowledge. And, as we have already seen, scientists like William Gilbert and even Isaac Newton were not immune to Hermetic mysticism.

But things were somewhat different across the Channel. Marian Mersenne, friend of Descartes and devout Catholic, was mounting a fierce attack on Neo-Platonism and Renaissance magic, picking out Robert Fludd for especial scorn as a pretended magus. Indeed, the whole spirit and practices of the Hermetic philosophy was deeply alarming the Church, especially alongside the advent of what Galileo called The New Science. Sorcerors were routinely burnt at the stake, and the practitioners of the New Science had to watch very carefully what they said, couching claims in terms of 'as if' rather than fact.

The New Science

Down the centuries from the time of Plotinus the Roman Church had become very powerful, spawning crusade after crusade against their enemies. Initially, these enemies, naturally enough, were the Saracens occupying the Holy Land. But the corruption of the Christian message, born out of fanaticism and greed, soon manifested itself in the brutal suppression of their erstwhile servants, the Knights Templar, and the 'crusade' against the Cathars, labelled heretics for their idiosyncratic Christian beliefs. Christian atrocities against Christians became official with the creation of the Inquisition, run by the Dominican monks. A geographical lessening of the Church's power came with the Reformation, but Man and Earth remained the centre of the world and of God's attention. Shifting these from their divinely ordered position was anathema. The heavens were eternal; no change was possible. Yet such a change, beyond all anticipation of the medieval Fathers, did come, but there was no readiness; there was no world beyond belief.

The heavens were the first to crumble. Circular motion was the perfect sort of motion, taught Aristotle, so astronomers felt that they had to see the orbits of the planets in terms of circles. The motion of the Moon, Sun and the five planets was extraordinarily well-documented, but it was clear that their motion around the earth was far from circular, particularly the orbit of Mars. In the year 140, Ptolomy of Alexandria achieved lasting fame by describing planetary motion with great precision using numerous epicycles, circles within circles. It was hardly likely that epicycles in profusion would be the correct description, even though it did use the blessed circle, but it did *save the phenomenon*. The problem, of course, was that, in fact, the planets and the Earth rotated about the Sun, but Aristarchus of Samos, whose bright idea it was, was ignored, partly because that did not fit in with the prevailing dogma but also because his predictions were not as accurate as Ptolomy's. In the event, and ironically, it was a canon of the Church that revised the idea of Aristarchus, one Nicolaus Copernicus (1473–1543) in 1510. Certainly conscious that displacing the Earth from its hallowed position at the centre of God's universe was not going to be popular with the ecclesiastical mind-set, even with 'as if' disclaimers, he did not publish his famous book until 1543, which turned out to be the year of his death. Letting the planets, including the Earth, move around the Sun simplified matters considerably. But, even so, Copernicus still needed thirty epicycles to get a decent fit. It was not until Johannes Kepler (1571–1630) deduced that a planetary orbit was elliptical and not circular that the Sun-centred system showed its real power.

The name of the young Kepler came to the attention of the famous Danish astronomer Tycho Brahe (1546–1601) through Kepler's book written in 1596, in which he proposed a distinctly mystical Neo-Pythagorean model to explain why there were only 6 planetary spheres. These corresponded to the orbits of Mercury, Venus, Earth, Mars, Jupiter and Saturn (Uranus, Neptune and Pluto were yet undiscovered). There were only six spheres, he argued, because there are only five regular solids, and these somehow determined the spaces between the spheres. After Tycho's death Kepler was appointed in his place with full access to the detailed data on planetary orbits patiently obtained by Tycho, particularly data on the orbit of Mars. In 1605 he deduced that the orbit of Mars was elliptical, and he subsequently announced his famous laws of planetary motion that in years to come Newton was to explain in terms of gravitational theory. By 1605 Kepler had forsaken mysticism for mathematics.

But few could accept Copernicus's idea that the Earth moved — wouldn't one feel something, a wind maybe? It was a truly shocking idea. But there was another shock to come. Almost thirty years after Copernicus's book was published, in 1572, a new star suddenly appeared in the constellation of Cassiopea, visible to all. Comets were relatively familiar, but a new star was unthinkable: the heavens did not change! That it was indeed a new star and not a comet was proved by Tycho Brahe's measurements over time; the star did not change its position with respect to the 'fixed' stars, so it was one of them. The concept of an unchanging heavenly firmament was shattered. We now know that the star was a supernova, an exploding star. They are rare, but Babylonian and Chinese astronomers must have been aware of them. If so, the knowledge did not travel.

Religious dogma was rattled, but the Church soon fought back. When in 1600 Giordano Bruno refused to recant his ideas about there being an infinity of worlds among the stars, he was burnt at the stake. When Galileo (1564–1642) insisted that the Earth *did* move, he was forced to recant, and was placed under house arrest in Florence. But all of this was too late. The Christian cosmology, and, indeed, the Neo-Platonic cosmology, was crumbling and giving way to the brave new world of Kepler and Galileo, where mathematics was displacing faith.

Endorsed in spades by Isaac Newton. The new mathematics of the calculus, attributed to the German philosopher Leibnitz as well as Newton, was a tremendous creation of the mind. Its role in all major advances that were to come was crucial. Its importance cannot be exaggerated. The challenge presented by the motion of things was to invent a mathematical way of describing speed at any given instant and how it changed with time. Speed was simply the distance travelled in a certain time—miles per hour, kilometres per second—whatever. Getting speed defined at a given instant meant reducing the time interval to be as tiny as possible, which, of course, would make the distance travelled in that tiny time interval very tiny as well, but such that that tiny distance divided by the tiny time interval was the speed. Imagining the time interval to be infinitesimally short then allowed us to define the speed as the speed at that instant. The invention of infinitesimals and their manipulation led to what we now describe as the calculus. It enabled theoreticians to describe the complex motion of matter under the action of forces by equations employing the neat language of the calculus.

And, in addition, Newton introduced that magical action-at-a-distance, namely, gravity, that was to be the despair of Descartes, who believed, as we have noted, that movement had to be initiated by palpable push-and-pull action, not by mystical forces that nobody could see. He had a point, but had he forgotten about the lodestone and amber? Claiming not to invent funny hypotheses, Newton showed that the concept of gravity plus the laws of motion discovered by Galileo accounted happily for the elliptical orbits and laws of planetary motion that Kepler had discovered. What more was there to be said! It certainly saved the phenomenon. The science of celestial mechanics was born, but at the expense of admitting the existence of a magical force.

There were more down-to earth changes affecting contemporary civilization. The art of medicine had been more or less dormant since the first century. Then the Roman, Galen, added his concept of animal spirits to the four bodily humours of the Greek Hippocrates viz: blood, phlegm, black bile (for melancholy) and yellow bile (for rage). And so medicine remained up to the sixteenth century when it began to be refined by the Swiss Paracelsus (1490–1541). Sporting the name Phillipus Aureolis Theophrastus Bombastus von Hohenheim, he was understandably happy to be known simply as Paracelsus. Reputed to carry pills in the hollow hilt of his sword, he was the first to apply chemicals to the treatment of diseases; and, with him, came some first awareness of the role of microbes in the spreading of infections. Visiting a patient, his opening remark was said to be 'I am different; let this not upset you.' A giant step forward was made in the following century by Harvey's discovery of the circulation of the blood. Alchemy continued its transformation into chemistry and in Agricola's classification of minerals the science of geology was heralded.

The study of crystals in their own right that was to lead to the science of crystallography began to get under way. For the mathematically minded, gone was the magic. Replacing it was the discovery that certain angles between crystal faces turned out to be more common than others. A Law of Nature that described these special angles in terms of whole numbers was discovered by a cleric, one Abbé Rene Hauy in 1792. This was a hint that crystals were composed of discrete entities—atoms—but the idea of atoms was still a long way from being accepted.

If the impression gained from famous persecutions that the Church was wholly against science, this would be a mistake. Any belief expressed that contradicted the Bible was, of course, unacceptable, but scientists became adept at describing their beliefs as 'as if' theories. The Earth was the centre of the world, but one could describe planetary motion as if the Earth were going around the Sun. In any case, many scientific discoveries were in areas where the Bible was silent. There was therefore nothing to stop the scientific cleric getting involved, and many famous scientists were churchmen. In those days, bright young men without a private income, but who were lucky enough to find a patron to point them in the direction of the Church, where they could continue their education, became clerics. So it is not surprising to find a sprinkling of Abbés and Doctors of Theology in the scientific pantheon.

The belief that matter was composed of atoms—discrete entities that could not be subdivided—an idea that goes back to the hypothesis of Leucippus and Democritus in the fifth century BC, had to fight the disapproval of that heavyweight, Aristotle, who did not like the idea of a void between the atoms. And nor did René Descartes nearly 2000 years later. Nevertheless, the idea was revived in a book published in 1649 by one of his fellow countrymen and contemporary Pierre Gassendi, a Doctor of Theology, no less. There was a distinct horror of the idea of—nothing. But if Aristotle and Descartes believed that 'Nature abhors a vacuum', this was being given the lie by a compatriot of Galileo, Evangelista Torricelli, who was the first to demonstrate that, if not a vacuum, then something pretty close to it could actually be created—a tube with the air sucked out, in what was essentially the first barometer. This business about the vacuum was going to come up time and time again in physics, and it is a lively topic today in connection with cosmological dark energy, but whatever people thought about it, it did not stop the idea of atoms taking hold.

One of the problems that confronted the anti-atomists was simply this. Given two lumps of matter of the same size but of different weights — say a lump of glass and a lump of gold — how to explain this without assuming that the lighter glass was emptier than the gold. As Newton had explained, more weight means more matter, so if matter is composed of atoms the atoms of gold must be individually heavier than those of glass, so if the lumps are of the same size glass must be emptier than gold. True, said the anti-atomists, but only if you believed in atoms. If matter is not discrete the matter that composes glass is just less dense, but it fills the whole space. No voids!

The debate about the existence of atoms rumbled on and on. Strong support for the atomic hypothesis continued to come from chemistry. But in spite of all the chemical evidence, the anti-atomists remained unconvinced, still worried about the void — worrying about nothing, as it were. It took Ernest Rutherford's demonstration in 1909 that alpha particles penetrate metal foils and scatter in a way that shows that matter really did have plenty of empty space. By which time (1896) Wilhelm Rontgen had discovered X-rays, electromagnetic waves that turned out to have wavelengths of the order of the size of an atom, and twenty years on the Braggs, father (William) and son (Lawrence), were busy studying how the lattice of atoms that made up a crystal could be quantified by X-ray diffraction. The science of crystallography had become a branch of physics. So, relax, it is no longer way out or eccentric to believe in the existence of atoms. The evidence for them is simply overwhelming. But are the gaps between atoms really empty? Watch this space.

X-ray diffraction confirmed to many that the regular shape of crystals was a consequence of crystals being composed of a regular array of atoms. Instead of the geometry being that of a Euclidean continuum, it was more Pythagorean, being idealizable in the simplest picture to a lattice of points. Ignoring the actual nature of the atoms or molecules that compose the crystal allows structure to exhibit properties of a purely mathematical kind. There were plenty of observations to be explained. The natural history of crystals had identified just seven crystal systems: cubic, hexagonal, tetragonal, orthorhombic, monoclinic and triclinic. Why seven? And then there are those angles that the Abbé Hauy had identified. How are these to be explained?

The answer to these problems is to be found in the symmetry properties of the crystal lattice. A perfect lattice is an extended structure that is made up of a building block of points repeated over and over again. The first requirement, therefore, is that the lattice must look the same if its building block is translated in space. Space being three-dimensional there will be three translations that define the translational symmetry. This means that there are three directions along which the lattice of points can be shifted by a certain distance, that depends in general on the direction of shift, without changing the look of the lattice. In a cubic crystal, for example, the directions will be at right angles to one another and the amount of shift will be the same for all three directions. Such directions define the crystal axes.

Another operation that can leave the lattice looking the same is a rotation. Such a rotation would explain how crystal faces come about. Now any rotation has to be such that translational symmetry is not violated. It is a fact of three-dimensional space that this condition reduces the number of possible rotations to just five! The allowed ones correspond to rotations of 60^0, 90^0, 120^0, 180^0 and, of course, 360^0. If a 60^0 rotation is allowed there can be six of them without changing the lattice. This is exactly the case for a hexagonal crystal, so the latter is said to have 6-fold symmetry. A 90^0 rotation implies a 4-fold symmetry, which is that of the cubic crystal. The other angles refer to 3-fold, 2-fold and 1-fold symmetries. We notice that the formula describing these angles is just $360/n$, where n is 1, 2, 3, 4, or 6. These exactly corresponded to the numbers discovered by Abbe Hauy. Crystals no longer had occult, magical properties! A victory over animism!

It was bad enough for the Church that the eternal nature of the heavens had been shown to be false; that the Earth had been demoted to the status of an ordinary planet; that everything was governed by mathematics; without Man himself becoming no longer a direct creation of God but merely a product of purposeless evolution. It is small wonder that Charles Darwin (1809–82) delayed publishing his Theory of Evolution; it was bound to be shocking. He was persuaded to publish only after receiving a letter from Alfred Wallace, who had been reaching similar conclusions. Darwin published *The Origin of Species* in 1859, and the world has not been the same since. His basic idea was that species had evolved over countless millions of years by natural selection, a combination of adaptation to the environment and the survival of the fittest. More fundamental evolutionists see life as having been generated out of matter in 'the primeval soup' of the cooling Earth four thousand millions of years ago with primitive life-forms evolving to produce bacteria, plants, insects, fish, mammals and Man. This view of Man being a product of a purposeless, mechanistic process naturally did not appeal to the religious then, and it certainly does not appeal to the religious now, who prefer the idea of Intelligent Design over Mindless Evolution.

The debate that Darwin's book has created is very much alive today, as we will see in a later chapter. It concerns life. The rest of the world had long been killed off and was being busily dissected. The patient experimental study of electricity and magnetism over the seventeenth, eighteenth and nineteenth centuries produced a wealth of detail that, in 1864, James Clerk Maxwell found could be summarized in just four equations. Expressed in the notation of the calculus, they described how electric and magnetic forces arrayed themselves in space, how magnetism was generated from electric currents, and how electricity was generated from changing magnetic fields. The magic of the lodestone and amber had been domesticated and any incipient animism destroyed.

But what constituted an electrical current? A fluid of some kind—maybe two sorts of fluid. The solution was found in the study of the flow of electricity in a partially evacuated tube in which a glow was seen to emanate from the cathode (the negative terminal). In 1879 William Crookes showed that these cathode rays consisted of particles that could be deflected by a magnetic field to produce a phosphorescent glow where they collided with the walls of the tube—a less sophisticated glow than that of your old TV set, but the same effect. In 1897, J. J. Thomson, studying the way in which these particles were deflected by a magnetic field, got a value of the ratio of the charge to the mass, e/m. Subsequent measurement of the tiny charge, e, by looking at how charged droplets of water moved in an electric field, enabled the mass to be deduced from the e/m measurement. The electron had been discovered. The glow could be understood in terms of fast electrons emitted from the cathode knocking electrons off the atoms of the residual gases in the tube, i.e., ionizing the atoms, and the electrons being recaptured by the atoms and emitting light in the process.

An electric current, then, was a flow of tiny particles called electrons, each carrying a tiny negative charge of electricity, each having a mass almost 2000 times less than a hydrogen atom. The hydrogen atom was now seen to consist of two particles, the proton, roughly 2000 times heavier than the electron but with a tiny positive charge of the same magnitude as charge on the electron. The attraction of the electron to the proton held the atom together. Static electricity arose from the removal or addition of electrons or ions, ions being atoms that have become charged by either acquiring or losing electrons. A metal wire carrying a current contained electrons flowing under the influence of an electric field through a background of positively charged ions.

Maxwell's theory did a more remarkable thing yet, it conclusively showed that light itself was nothing more than an electromagnetic wave whose speed in a vacuum was a fundamental constant. For any velocity to be a fundamental constant was shocking. To the whole of physics devoted to the motion of matter, the classical mechanics initiated by Galileo and Newton and developed into a wonderfully comprehensive mathematical system by, *inter alia*, Lagrange, Euler, Laplace and Hamilton, motion was always motion relative to a fixed background. But here was Maxwell proving that the velocity of light was somehow an absolute velocity. This was too much, and so was born the concept of an invisible, all-pervading aether that supported light waves whose velocity was relative to it. In the fourteenth century, William of Occam (a turbulent English priest) enunciated a general principle to guide those intent on describing things: 'It is vain to do with more what can be done with less.' Occam's Razor was to appeal to Ernst Mach (1838–1916, famous for supersonic Mach numbers) who advocated searching for 'the simplest and most economical abstract expression of facts'. It was to lead to the death of the concept of the aether. That the velocity of light was *not relative to anything*, as all other velocities were, was accepted by Albert Einstein, and it became the corner-stone of his Theory of Relativity, which was to change our understanding of time and space.

The message of Classical Physics — electromagnetism and mechanics — that invaded European culture could not have been more inimical to the Magical World. Given knowledge of the starting conditions of the motion of matter, never mind what sort of matter, all subsequent evolution was absolutely determined and predictable. Laplace formulated it with precision.

> An intellect which at a given instant knew all the forces with which nature is animated, and the respective situations of the beings that compose nature — supposing the said intellect were vast enough to subject these data to analysis — would

embrace in the same formula the motion of the greatest bodies in the universe and those of the slightest atom: nothing would be uncertain for it, and the future, like the past, would be present to its eyes.

If the system was very complex, such as an organic living thing then, of course, any prediction would be extremely difficult, but *in principle* doable. Thomas Huxley in his enthusiasm for Darwin's theory of evolution had no doubts.

If the fundamental proposition of evolution is true, that the entire world, living and not living, is the result of the mutual interaction, according to definite laws, of the forces possessed by the molecules of which the primitive nebulosity of the universe was composed, it is no less certain that the existing world lay, potentially, in the cosmic vapour, and that a sufficient intellect could, from a knowledge of the properties of the molecules of that vapour, have predicted, say the state of the Fauna of Great Britain in 1869, with as much certainty as one can say what will happen to the vapour of breath on a cold winter's day.

In this brave new world mechanism replaced magic; atoms existed; the void existed. The world was henceforth to be viewed according to the Doctrine of Scientific Materialism. Outside of religion, the soul became an ignorable efflorescence of the material brain.

And here we reach the spiritual nadir created by the materialist world-picture. Classical mechanics claimed to have proved that everything is determined by impersonal forces: free will is an illusion, there is no God, religion is a myth, morality is arbitrary.

This scientific nihilism destroyed the idea of human life having any meaning whatsoever. If this were all that science could contribute, things would be dire indeed. Fortunately, most individuals either retain a stubborn scepticism in the face of this dogma, or ignore it altogether, and get on with their lives. And here, science could make a substantial contribution to the material welfare of the peoples of Europe and

North America. In doing so it realized the vision of Francis Bacon to be, first and foremost, of benefit to mankind. This was a substantial achievement to be set on the scales to balance, or at least counterweight, its materialistic message. The brute material power that science began to exhibit in the nineteenth century forged, not only a base for our contemporary affluence, but also it gained a popular prestige far beyond its intellectual importance. We turn in the next chapter to a brief account of some of those benefits.

Chapter Five

For the Good of Humanity

What motivated all those scientists, spread over Europe and the United States, to search for and discover the laws that govern the behaviour of matter? Was it a dispassionate study of Nature that forged a revolution, a revolution in the way everyone perceived the world about them? Though this is a view, commonly encountered, it is far from the reality; there were many strands, some of them far from being notable for their coolness and lack of irrelevant passion. For some, like Blaise Pascal and Isaac Newton, there were strong spiritual elements in their investigations. In his *Pensées*, Pascal affirms a view that science is the proper study of God's world, a form of worship, to be done seriously and piously. In works much less famous than his monumental *Principia*, Newton betrays an altogether darker side, more recognisable as that of the renaissance magus than that of the sober physicist. What is revealed is a Faustian appetite for arcane knowledge wherever it may be found — in the stars, in the furnaces of the alchemists, in the Bible, in prophesies — even in the mysteries of the natural magic of the Hermetic tradition. For many of their contemporaries there was a deeply religious aspect to their science, and many of those that followed, down to the present day, retained that disposition. For others of a colder intellect there was, indeed, only that dispassionate curiosity about Nature that one expects of a paradigmatic scientist.

But whatever the cast of mind, the challenge to create a comprehensive understanding of the universe was a shared aim.

Personal motivations differ, but in the sixteenth and seventeenth centuries when it all began, two complementary strands were evident; the one, hot and passionate, personified by the German, John Faustus, the other, cooler and practical, by England's Lord Chancellor, Francis Bacon (1561–1626). Bacon is widely regarded as the founding philosopher of science, with a distinctly more humanist bent. For him, the power over Nature was not to be sought by invoking demons and spirits, but rather by patient and systematic observation leading to the induction of certain laws. In his *New Atlantis* he described what a research laboratory might be like, and, certainly, it was highly advisable to keep God out of such a laboratory, and that went for all laboratories! There is the story of the famous French mathematician, Laplace, explaining his mathematical theories to Napoleon who pointed out that God did not appear in his scheme, to which Laplace replied that he had no need of that hypothesis. But Lagrange, (another famous French mathematician) who was present, murmured that yet God could explain much! Which was exactly why Bacon, pious as he was, saw that God ought not to replace a scientific explanation.

Above all, science, godless though it had to be, was to be for the good of humanity. Bacon's whole ethos was echoed all over Europe. Societies, formerly 'invisible' for prudent reasons connected with the Church, became public. One of the first, the Royal Society, came into being in 1660 with the blessing of the newly restored King Charles II, dedicated, as it is today, to 'the improvement of natural knowledge'. A powerful movement throughout Europe had appeared, but it was not without its critics. One of these, the Irish Dean Swift, lampooned the Royal Society by having his protagonist in *Gulliver's Travels* visit the floating island of Laputa, where were found mumbling scholars and idiot schemes, like trying to get sunbeams out of a cucumber. (Actually, these days the scheme is not so idiotic: even a cucumber could be induced to lase and produce its own characteristic beam, if not of sunlight.) Voltaire's comments concerning these Academies includes:

> The Royal Society of London lacks the two things most necessary to man: rewards and riches. In Paris membership of the Academy means a guaranteed small fortune for a mathematician or a chemist; on the contrary, it costs money to belong to the Royal Society.

But what none of the early founders of science perceived in any detail is the down-to-earth fuelling of science by industry and business. Science was motivated by the quasi-religious desire to understand Nature: the irony is that it has mostly been the practical needs of humanity that have motivated science. The need for a means of travelling that was quicker than coach and horses led to the steam engine which, in turn, led to the birth of the science of thermodynamics. It stimulated James Joule, son of a wealthy brewer in Manchester, to an understanding of the relation between mechanical work, heat and energy, the Frenchman Sadi Carnot to his reversible cycle compression and rarefaction, and Rudolph Clausius to the concept of entropy and the second law of thermodynamics. The Industrial Revolution in England, based on weaving and initiated by the advent of cotton from the New World, needed machines that accelerated production by reducing time spent in laborious manual tasks. One response, the use of cards bearing a pattern of holes, was invented to simulate repetitive tasks in weaving, an early form of a task-dedicated computer.

Communication between the cotton plantations in the southern United States and the factories of Lancashire was, naturally, an important factor, and it became highly desirable that this be made as rapid as possible. The need was felt by business generally. A merchant had to endure a wait of weeks, sometimes months, before learning how successful his ship had been. Faster ships would help, but faster communication would be even better. The same need was just as urgent for the military. The result of a battle fought far away could take weeks to reach home. In the spring and summer of 1805, all England was fearful of an invasion across the channel by Napoleon's crack army. There were only a few English ships to deter an attempt, but they did; Napoleon shifted his attention to the East and by the 20 October was receiving the surrender of the Austrian army at Ulm. Horatio Nelson's victory over the French and Spanish fleets at Trafalgar on the 21 October effectively eliminated any future threat, but news of the victory did not reach England until the 4 November, two weeks later. A further consequence of Nelson's victory was that England became the supreme sea power, which allowed her to extend her empire rapidly. With a global empire to guard, the military felt the need for rapid communication even more keenly.

At the time, telegraphy was confined to line-of-sight beacons on high hills and semaphore with flags. Sir Charles Wheatstone (1802–75, famous for inventing the rheostat and the Wheatstone Bridge, which school children used to come across in the physics lab), conceived the idea of transmitting messages by sound vibrations along a rod. He estimated that sound would travel at about 200 miles a second, and he proposed a telegraph, based on this idea, to be set up between London and Edinburgh. This came to nothing, but science had already begun to explore the mysteries of electricity and magnetism, and aspects of this new knowledge began to look promising. By the end of the eighteenth century enough had been learnt for science to follow Bacon's ideal and make a contribution to the well-being of humanity. André Marie Ampère's study of the interaction of an electric current and a magnetic field became the basis of a number of novel devices, among them the electromagnetic relay that was to play an important part in a number of systems. It was found that a current passed through a coil of wire could deflect a compass needle. The electrical engineer, Sir William Fothergill Cooke saw that this effect could be the basis of an electric telegraph and, having persuaded Wheatstone, Cooke and Wheatstone produced the first working device in 1837. It consisted of five needles that could be stimulated by an electrical signal to rotate and indicate a letter of the alphabet. Wheatstone also measured the speed at which an electrical signal propagated down a wire and found it to be near the speed of light (186,000 miles per second). Later, Michael Faraday (1791–1867), a close friend of Wheatstone's, showed that the speed of a signal propagating along a submarine cable was somewhat less, being 144,000 miles per second at most. Nevertheless, the speed of an electrical signal in a wire was fast enough to satisfy everybody.

In the United States a certain Professor of Painting and Sculpture, one Samuel Finley Breeze Morse (1791–1872), hearing about electromagnets, conceived that these could be exploited in telegraphy. The electromagnetic relay was nothing more than a switch, but it stimulated the invention of electrical telegraphy by Morse. Never quite abandoning art, he devoted most of his time in developing his idea, and in 1844 he sent his first message in a code in the form of a series of dots and dashes—the Morse code—between the Supreme Court Room in the Capitol in Washington to Baltimore. The message was rather gnomic 'What hath God wrought.' Subsequently, telegraphists working Morse keys ran the risk of developing what is now known as repetitive strain injury (one thing that God wrought?).

The invention of the electric telegraph was a huge step forward and boon to the railway system. In fact the railways were the obvious tracks by the side of which to run telegraph wires. But wires were needed. This was not an insuperable problem on land, but getting a cable across a body of water was a challenge. The first submarine cable was laid on the sea bed in 1851 across the English Channel between Dover and Calais, soon to be followed by others across the seas of Europe. A substantially greater challenge was to make contact with America, but it was eventually done. One of the great achievements of Lord Kelvin, otherwise known in connection with the scale of absolute temperature, was to mastermind the laying of the first Transatlantic cable in 1866. The invention of the telephone followed closely. The first patent went to Alexander Graham Bell (1847–1922) in 1876, which laid the foundation of the Bell System Corporation in New Jersey. Telephony had been born; one could now listen to someone talking no matter how far away, provided there were suitable cables in place.

Faraday's investigation of electromagnetic induction stimulated a host of new devices, not least of which was the electric dynamo and the electric motor. His work also stimulated James Clerk Maxwell's (1831–79) Theory of Electromagnetism. Maxwell's prediction of electromagnetic waves in 1870 and their discovery by Heinrich Herz in 1886 heralded a new phase of telegraphy. Electromagnetic waves made wireless communication a possibility, a possibility that was exploited by an Italian, Guglielmo Marconi (1874–1937), who invented antennae to transmit and receive radio signals. In 1896 he travelled to England and demonstrated his apparatus and received the world's first patent for wireless telegraphy in that year. The next year he formed a company which in 1900 became Marconi's Wireless Telegraph Company Ltd and, soon, simply Marconi. In 1909 he was awarded the Nobel Prize for Physics. The principle was simple—an oscillating electric current in an aerial radiated waves that could be detected at a distance; impressing a message on the waves meant that wireless communication could be achieved with a speed limited only by the speed of light. Bouncing waves of sufficient power to overcome inevitable absorption and scattering losses off the ionosphere meant that communication could be global, though, as anyone who has tried to listen to short- wave broadcasts will agree, the reception can leave a lot to be desired. The company that Marconi founded, and the last British company to manufacture telecommunication equipment, was eventually taken over by the Swedish company Ericsson in 2005.

Speed was now no longer the problem. Electric telegraphy and radio transmission had reached a natural limit of speed set by the velocity of light. The need now was for clarity of transmission and the ability to transmit large amounts of information along a single cable or via a single antenna. This was the motivation for further technological advances, and it still is.

Government in all its aspects, industry, banks, business at large, and also the military, employ clerks many of whom, and perhaps all of them at one time or another, are occupied in adding and subtracting figures and generally manipulating numbers. This can be a time-consuming activity in commercial activities, and, in government, whenever there is a census. The Hindu invention of zero and the adoption of the Arabic numerals in the base 10 was a huge improvement over the Roman scheme, but troubles involved in representing decimals had to wait till the eighteenth century before the decimal point was generally accepted. Fingers and toes were useful aids—still are—and there was also the abacus, but new ideas to speed calculations up and make them more accurate were sorely needed, especially since most clerks were innumerate, and since trade and commerce were expanding. Beyond a certain point memorising more and more multiplication tables is not practicable, and long multiplication and long division was slow, tedious and often wrong.

John Napier's (1550–1617) discovery (or invention, for those not persuaded of the existence of Platonic forms!) of the logarithm alleviated these difficulties of multiplication and division. By expressing a number as a power of 10, i.e., its logarithm, multiplication was reduced to the addition of exponents and division to the subtraction, and the result converted back to a number, i.e., its anti-logarithm. With tables of 'logs' and 'antilogs' available, multiplication and division became less of a chore. Much later, the idea of logarithms led to the representation of a number by a length of a line along a ruler so multiplication and division could be done simply by adding or subtracting two lengths. This is the basis of the slide-rule, a device that was standard equipment until the pocket calculator came along.

With the appearance of machines capable of reducing labour in the weaving industry, it was natural to wonder whether a similar reduction of labour might be possible by some sort of calculating machine. The first of a long line was the Pascaline, an invention of Blaise Pascal in the seventeenth century, which consisted of a set of interlocking cogs and wheels. A wheel with ten teeth could handle the digits from 1 to 10; a second wheel attached to the first so it rotated one step whenever the first wheel got to 10 and this 'tens' wheel was similarly attached to a 'hundreds' wheel, and so on. Dialling the numbers to be added just rotated the wheels appropriately and at the end the result was automatically displayed. This made addition much easier, but it was certainly not quicker. Multiplication, which is just repetitive adding, was very laborious, and it was only when Gottfried Leibnitz, famous among other things for his development of calculus, improved the design by adding a multiplier wheel, one with nine teeth of different lengths, that a significant speeding up was achieved.

But it was not until the nineteenth century that the first computer was built. The designer was Charles Babbage (1792–1871), whose machine was called the Difference Engine. This was an immensely complicated structure consisting, once again, of cogs and wheels, designed specifically to solve polynomial equations. In 1828 he was appointed to the Lucasian Chair of Mathematics at Cambridge and in 1831 he founded the British Association for the Advancement of Science. He was something of an eccentric—at one stage even stood for parliament as a Whig (unsuccessfully). He was not very successful at persuading funds to flow his way in order to develop his ideas about calculating machines further. (Oddly, for a mathematician, he disliked music.) But he was convinced that a more powerful machine than his Difference Engine could be made. Calculating polynomials was seen by Babbage to be too limited: what was needed was a machine that could tackle any problem. This called for a much more ambitious structure which Babbage called his Analytical Engine, which was essentially the first programmable computer in that its parts were broadly recognisable as those in modern machines. It had an input section for feeding in numbers and instructions; it had an arithmetical unit that performed the calculations; it had a control unit that made sure that the calculations were carried out in the right order; it had a memory in which numbers were stored; and finally it had an output section that displayed the result. And all of this composed of cogs and wheels, gears and levers, and connections galore that needed to be turned to the greatest possible accuracy in order to avoid slackness and, in extreme cases, jamming. He also saw that the clever card with its special pattern of holes, used in weaving, could be pressed into service to control the sequence of calculations.

Babbage himself, his ambitions and the weaving connection made a great impression on a lady who was to be well-known through her writings about Babbage's Engine. The lady was the daughter of Lord Byron, the Countess of Lovelace (1815–52). Ada Byron was the fruit of the year-long marriage between Lord Byron, the poet, and a mathematically minded lady called Anne Millbanke. Having survived her mother's harsh regime of her education in the hands of tutors appointed more for their expertise in mathematics than in poetry or music, Ada grew up into a very personable and extremely bright young woman, married William King, Earl of Lovelace, and had three children. As Lady Lovelace she sought the society of the intellectuals of the day, among them Sir Charles Wheatstone, Michael Faraday, Charles Dickens, and, above all, Charles Babbage. An Italian, Menabrae had written a memoir on Babbage's Analytical Engine which Lady Lovelace, now fascinated by the concept of computational machines, wrote a translation, and then, with the encouragement of Babbage, added copious notes and annotations. In those notes appeared remarks concerning the power and possible intelligence of computers, remarks that inititiated a debate that, even now, rages today in the world of Artificial Intelligence. For that she is justly famous.

In spite of the brilliant ideas that led to the Analytical Engine, it became clear that a purely mechanical computer was too slow and too difficult to make reliable. Something new was needed. And here, once again, science came to the aid of engineering in the form of electromagnetism, as it had, in fact, done in responding to the need for faster communication. The new device was the electromagnetic relay. This does nothing more than switch a current on or off. Further developments led to the invention of the thermionic valve, which could also act as a switch. Which should have suggested something obvious to putative computing engineers, but it did not, except to a German engineer called Zuse.

The tradition, centuries old, was to handle numbers in the decimal system. It meant having ten different symbols — 0, 1, 2, 3, 4, 5, 6, 7, 8, 9 — and in Babbage's Analytical Engine it called for ten cogs on each wheel. Lady Lovelace's notes were read with great interest in two continents and in the first three decades of the twentieth century the development of calculating machines continued, particularly in the United States, Britain and Germany. The onset of the Second World War gave that development a priority which certainly bore fruit. In the USA the International Business Machine Company (IBM) had achieved considerable prominence, and it had produced a monster computer — Harvard Mark1 — which was the inspiration of a Harvard academic called Aiken. Its basic unit was the electromagnetic relay. In 1940 a team was set up in Bletchley Park, Hertfordshire, to crack the codes of the German military. Here the first electronic computer was built — electronic because instead of relays it employed thermionic valves. Relays operated by switching, at best, ten times a second whereas valves could switch at over a thousand times a second. Their machine — the Colossus — used an input of punched paper tape that could scan two thousand characters a second, which was a world record at the time. In Germany Zuse and colleagues were also using thermionic valves. Zuse was the first to realise that the decimal system was not ideal in an engine that used switches. The basic function of a switch is to turn a current on or off. In a system with only two characters instead of ten, one of them could be the on state, the other the off state. Such a system was well known to mathematicians — it was the binary system, in which each decimal number could be represented by a sequence of zeros and ones. Thus 0, 1, 2, 3, 4, 5, 6, 7, 8, 9, 10 in binary notation are 0, 1, 10, 11, 100, 101,110, 111, 1000, 1001, 1010, in effect, a series of ons and offs. Nowadays, the binary system is used in all computers. It was fortunate for Britain that the German Government did not support the construction of the computer that Zuse had designed. Its power could have meant that the military codes of the Allies could have been broken more quickly, and it might have been influential in the race to produce the atomic bomb.

The needs of industry, business and the military have continued to encourage science to extend its interest into the means for improving communication and computation and, for the most, this is proceeding with Baconian dispassion, but not everywhere. In her notes, Lady Lovelace remarked that a computer, though eventually capable of achieving much more than mere arithmetical functions, was, nevertheless, just a machine that obeyed instructions with no powers to originate anything. A computer that could handle any kind of calculation was shown to be valid by Alan Turing (1912–54), one of the Bletchley Park team, and his concept of a universal computer was seminal in the creation of a science for computing. He revived the question that Lady Lovelace had effectively raised long ago: could machines be regarded as having intelligence? He imagined a conversation carried out with a hidden person or machine: if one could not tell the difference, then, to all intents and purposes, the machine can think. Even if a machine could pass the Turing Test it was not obvious that its potential abilities were not limitless. Indeed, in 1931, the young Austrian mathematician, Kurt Gödel (1906–78) proved a theorem that, in the context of the present topic, meant that at any time there existed mathematical truths that could not be deduced from the existing set of axioms. Turing saw that this meant that, no matter how powerful a computer was, there were tasks that it could not tackle, whereas it is possible for a human mind to do so. For some in the field of Artificial Intelligence today there is a Faustian belief that a fully intelligent computer can be built. The debate triggered by the Lady Lovelace is highly relevant today and frequently fierce, especially as, these days, computers can play games like chess and win against experts.

But at the start of the twentieth century research into the design of computational machines, and what is now known as their architecture, had just begun. While new ideas in this area could indeed contribute to increasing the speed of calculations, the emphasis had to be on increasing the speed of the basic devices, the basic switches. Society's need for faster and faster computers became a powerful driving force for what may be termed the crystal revolution that was at least equal to that stemming from the need for faster, clearer and denser communication of information. Silicon, an abundant element, fashioned into a transistor, proved to be the ideal crystal to provide a fast alternative to the thermionic valve. On the way to satisfying the needs of speed and clarity, computer science was born, along with the sciences of electronics, opto-electronics and the physics of semiconductors. But what was first needed was new physics.

Chapter Six

The New Philosophy

In this new science three things have been forgotten. In all this, there is no mention of mind, no mention of soul, no mention of God. Science has no need of these hypotheses. Napoleon would not have been the first to be confronted by undiluted Laplacian materialism. The message permeating European culture was nakedly anti-religious and still remains so. If the world can be explained so vividly, so brilliantly and so powerfully, by the finest minds, who needs the Church with its unverifiable, unfalsifiable, message? Most of us in the West have comfortable, well-regulated lives. Food is plentiful, entertainment is plentiful, everybody has a TV and a mobile phone, most have a car, we can fly. Who needs God?

Plato needed God to provide a shared transcendentalism for the inhabitants of his Republic to give it strength. Kant, having comprehensibly refuted all philosophic proofs of God, found that a God was necessary to solve the problem of morality. In his *Critique of Practical Reason* he says: 'Two things fill the mind with ever new and increasing admiration and awe, the oftener and more steadily we reflect on them: the starry heavens above and the moral law within me.'

Nietzsche, very aware of the vacuum left by the death of God, painted a dispiriting picture in *Zarathustra*: no more noble virtues—'self-surveillance, self-discipline, self-overcoming, discipline in matters of the spirit, to control one's Pro and Con, to see life in all its joy and horror and still affirm life' — and all this without the ideal of God or anything transcendental. What do we become?

Behold! I shall show you the Last Man.

'What is love? What is creation? What is longing? What is a star?' Thus asks the Last Man and blinks.

The earth has become small and upon it hops the Last Man, who makes everything small. His race is inexterminable as the flea: the Last Man lives longest.

'We have discovered happiness,' say the Last Men and blink.

They have left the places where living is hard: for one needs warmth. One still loves one's neighbour and rubs against him for one needs warmth.

Sickness and mistrust count as sins with them: one should go about warily. He is a fool who stumbles over stones and men!

A little poison now and then: that produces pleasant dreams. And a lot of poison at last, for a pleasant death.

They still work, for work is entertainment. But they take care that entertainment does not exhaust them.

Nobody grows rich or poor any more: both are too much of a burden. Who still wants to rule? Who obey? Both are too much of a burden.

No herdsman and one herd. Everyone wants the same thing, everyone is the same: whoever thinks otherwise goes voluntarily to the madhouse.

'Formerly all the world was mad,' say the most acute of them and blink.

They are clever and know everything that has ever happened: so there is no end to their mockery. They still quarrel, but soon they make up — otherwise indigestion would result.

They have their little pleasure for the day and their little pleasure for the night: but they respect their health.

'We have discovered happiness,' say the Last Men and blink.

God never was in the remit of science, so logically could not be touched by science. Indeed, there is absolutely no logical reason for a belief in God to be remotely affected by the progress of science. Beliefs in the trappings of religion—the creation, miracles, incarnation—are other matters, but the core belief in the existence of a Superhuman Being cannot possibly be weakened by reason. One either believes or one does not. For many, a belief gives meaning and comfort to their lives. What else could give meaning?

Soul is also not part of the remit of science, but mind definitely is. I know I have a mind that is different from my body. I believe that other people have minds. I believe that mind is as ubiquitous and as real an entity in this world as any lump of matter. Anyone interested in acquiring an understanding of Nature cannot limit himself to the study of lumps of matter, he has to try and understand the relationship between mind and matter. Unfortunately, the scientific method, so outstandingly successful in illuminating the nature of matter, is ill-adapted to the study of the mind. One obvious problem is that each of us is unique and science simply cannot deal with the unique. In cosmology, for example, it cannot deal comprehensibly with the unique universe we live in but has to invent the existence of multiple universes so that their properties can be treated statistically. But, quite apart from the business of uniqueness, where is the scientific language to describe mental life—love of music, grief, religious experience? Faced with impotence, science has to either yield to art, to literature, to music, to portray an understanding of the mind, or it has to deny that any problem exists. In the latter case it is claimed that once the neurology of the brain is fully understood, or, alternatively, once a suitable powerful computer is built that exhibits mind, the problem will be solved. One feels that there is a vast category error here!

A New Philosophy in the seventeenth and eighteenth centuries attempted to understand what was going on and what the New Science was about. René Descartes (1596–1650) was the first to put the 'I' into the debate about the nature of things. His method was to doubt everything except the fact that he thought. *Cogito ergo sum*, I think therefore I am. Whatever may be the world outside of my body, I think and am aware of thinking. Surely this is a good place to start — we are all aware of an 'I' that thinks. The mind comes first, sensations that the mind experiences arising from the body or the world outside come second. There is a clear duality between mind and matter in the world. The problem is how to reconcile them. An immediate conflict with the New Science arose. I make the simple act of moving my arm. Where does this motion come from and where does the energy come from? The act seems to ride roughshod over the laws on conservation of motion and energy. Descartes invokes the hand of God at this point, and makes no apology since his idea of God has more reality for him than any external finite substance. In his Third Meditation he describes the duality between mind and matter in terms of those classical ideas of substance.

> For when I think that a stone is a substance, or a thing which is capable of existing of itself, that I also am a substance, although I see clearly that I am a thinking and non-extended thing, and that the stone, on the contrary, is an extended non-thinking thing, and that there is a notable difference between these two concepts, all the same they seem to come together in that they both represent substances.

In giving priority to mind, Descartes opens the way to a subjective idealistic philosophy: by regarding mind as substance, he opens the way to materialism! But he is always clear about the distinction between mind and body: 'I remark

The New Philosophy

... that body, by its nature, is always divisible and that the mind is entirely indivisible.'

Matter always has the qualities of extension, figure, position and motion. God has been introduced. What of Him?

> By the name of God I understand an infinite substance, eternal, immutable, independent, omniscient, omnipotent, and by which I mean all other things which exist (if it be true that any such exist) have been created and produced.

Whatever one may think of substantive dualism, the philosophy of Descartes was seminal. It had an immediate influence on the continental philosophers Spinoza and Leibnitz, but in down-to-earth pragmatic Britain it evoked a cooler response.

For Descartes mind and matter had to be distinct substances. The plurality of nature had been reduced to a duality. But mind and body interact. Given what was meant by a rigorous definition of substance, Descartes's duality proved to be logically incoherent. So claimed his arch critic.

That critic was Baruch Spinoza (1632–77). Inspired by his notion of God and a deep devotion to logical and mathematical truth, Spinoza, around 1666, created a remarkable metaphysics that was beautiful if iconoclastic and at the same time one that embraced both science and religion. There was nothing like it before and there has been nothing like it since. At a stroke it banished the transcendental God of the three main religions, it demolished the Cartesian mind/matter duality and it described nature in a way that much modern science can subscribe to. These are extraordinary achievements. How did they come about?

The inspiration came from the Greek concept of substance—the essence of a thing. In ordinary usage we call wood a substance, iron a substance, and so on, but this is not what the Greek philosophers and Spinoza meant by substance. For them the substance of a being was its essence, the thing in itself. And, as Parmenides had made clear, what *is* now cannot become so in the future, nor can it pass away. If essence is to mean anything real, it has to be changeless. In whatever way substance is defined, it must be eternal.

The definition that Spinoza gives in his *Ethics* runs as follows: 'I understand *substance (substancia)* to be that which is itself and is conceived through itself: I mean that, the conception of which does not depend on the conception of another thing from which it must be formed.'

This definition distances philosopher's substance from ordinary usage. A modern way of seeing that wood, for example, is not a substance in Spinoza's sense is to recall that it is composed of molecules arranged in a certain way so there are conceptions here that bear on the nature of wood that have nothing to do with wood *per se*. Similarly, iron is not a substance since it is composed of atoms with magnetic properties, the concepts of atom and magnetism being far more fundamental than iron.

Of course, Spinoza did not have the discoveries of modern physics to call upon, but nor did he need them. His aim, one shared with Descartes, was certain knowledge, but whereas Descartes was satisfied by his *cogito ergo sum*, Spinoza aimed for the sort of certainty that mathematics affords, such as knowing that the angles of a (Euclidean) triangle add up to 180^0, or that the missing number in the series 1, 2, 3, 4, 6, is 5. He found that certainty in logic.

So if wood and iron were not substances, what were they? To Spinoza they were simply modes of appearance, as were every one of the myriad things and events of nature. Modes were thoughts, ideas, people, stars, planets, all were forms —essentially transient forms—of the underlying reality. Given the bewildering panoply of modes of appearance it was the task of the intellect to detect what were the attributes of substance. Spinoza noted that there is no reason to stop a substance having more than one attribute, so the intellect must be prepared to comprehend a substance in a variety of ways. The question then is how many fundamental substances are there? The Pre-Socratic Greeks postulated the existence of one, but can that be justified? Spinoza emphatically showed that it can.

The argument runs along the following lines. Let us suppose there to be twenty fundamental substances. If so, there must be a reason why twenty and no more or no less. Whatever the reason is, each of the twenty substances must contain this attribute of 'twentiness' which is something lying outside their natures. By definition, therefore, this common attribute of twentiness rules out each member as a candidate to be a substance. The same argument can be applied to any number greater than one, with the same conclusion. Given that definition of a substance, there can be only one.

The most striking result of this logic is the demolition of the Cartesian two substances of mind and matter. In Spinozan metaphysics thought and extension are two attributes of a single substance. This is his famous double-aspect theory of mind and matter. The fundamental substance that is Nature is both mind and matter. It is difficult for our poor intellect to comprehend that thinking stuff and extended stuff are one and the same substance. We have to abandon ideas about how mind is related to body because mind *is* body. What goes on in the mind/body is experienced as the aspect of thought or as the aspect of extension.

Furthermore, it follows that substance, being that it is, cannot be limited in time and space. Since substance cannot be created by another substance, existence has to be part of its essence. Any limitation implies the existence of some other substance. But no two substances can share the same attribute, in this case the attribute of existence, so there can be no limitation. Substance is therefore eternal and infinite. Spinoza proceeds to envisage substance having an infinite number of attributes, mind, matter and existence being only three of them. It is then natural for him to equate substance with God who, therefore, necessarily exists.

It is obvious the God that Spinoza envisages is vastly different from Jehovah, the Trinity and Allah, each of which inhabits a transcendental world remote from the natural world. The Spinozan God is in the thick of it, as immanent as can be. God is Nature itself, existing everywhere. In the old religions, God is a noun. In Spinoza's pantheistic world, God is a verb. God is involved in every mode of mental and physical behaviour. He is mind *and* matter.

So, what about man? Human beings are finite modes in Nature, to be conceived either under the attribute of mind or under the attribute of extension. Like all individual things they endeavour to retain their identity, and such endeavour is their essence. The difference between animate and inanimate things is that animate things are more complex and so can cope with a great variety of change without losing their identity. Such successful self-maintenance is clearly a property of life. A human body is affected by all manner of external things and so therefore is the mind. The identity of mind and body in man is just a special case. In fact, all things are an identity of idea and extension. In God, in Nature, there cannot be an extended thing without its idea, nor an idea without its extended thing. The idea here is essentially the Platonic one. In ordinary usage idea is not what is meant here. We might have an idea that a unicorn exists, but it doesn't mean that because we have that idea then a unicorn certainly exists. Spinoza distinguishes between adequate and inadequate ideas, an adequate idea being one that corresponds to an attribute of reality and is therefore absolutely true, whereas an inadequate idea is one that does not relate logically to other judgements. The relative truth or falsity of our ideas is to be judged by how well they logically cohere with the most coherent system of ideas about the world we have.

Gottfried Wilhelm Leibnitz (1646–1716)was a poly-math — scientist, mathematician, moralist, philosopher and much else. His notation of the infinitesimal calculus was somewhat neater than Newton's, and is the one mostly used today. His intimate understanding of the New Science was unrivalled. What disturbed him was its unremitting mechanical nature that failed to take into account the organic, purposeful features of life and thought. It is a criticism that is as valid today as it ever was. He sought for a unifying principle that would encompass physical, biological and psychological phenomena, and he found it in the concept of the *Monad*.

His thinking was in many ways influenced by Aristotle. The void did not exist, the soul was the form of the body, the hierarchy of the Chain of Being was continuous and infinitesimally graded. To define matter as that which is extended, as Descartes did, failed to recognize the details of the internal structure of things, especially the internal structure of living things. The idea of atoms and the void was not adequate. Instead, Leibnitz filled the world with monads, each a dynamic centre of energy, spiritual rather than physical, with aggregations of them making up the dynamic bodies in space.

So what is a monad? It is, first of all, a divine entity, without extension, figure or divisibility. Not easy to envisage, therefore. A monad is of itself, unaffected by anything, but mirroring the whole universe. It is part of a vast, continuous hierarchy, with God as the greatest monad, that stretched down to the least. The least experienced only perception; those monads that experienced both perception and desire were souls. If the monads did not interact, how was the world to function? Here, Leibnitz introduces the idea of a pre-established harmony that God had established, so that each monad feels all that is happening throughout the universe. Out of all possible universes, God has chosen this one as the best of all possible worlds.

Monads are aware of everywhere in the universe because the universe is a plenum.

> For since the world is a plenum, rendering all matter connected, and since in a plenum every motion has some effect on distant bodies in proportion to their distance, so that each body is affected not only by those in contact with it, and feels in some way all that happens to them, but also by their means is affected by those which are in contact with the former with which it itself is in immediate contact, it follows that this intercommunication extends to any distance whatever.

> And consequently, each body feels all that is happening in the universe, so that he who sees all, might read in each that which happens everywhere, and even that which has been or shall be, discovering in the present that which is removed in time as well as space.

Leibnitz's pre-established harmony makes everything determined, so past and future are there in the present. 'According to this system (pre-established harmony) bodies act as if ... there were no souls, and that souls act as if there were no bodies, and that both act as if each influence the other.'

He was very impressed by the wealth of living activity in a drop of water when observed through a microscope.

> Whence we see that there is a world of creatures, of living beings, of animals, of entelechies, of souls, in the smallest piece of matter.
>
> Each organic body of a living being is a kind of divine machine.
>
> It is a magnificent system, reminiscent in some features of the Hermetic mind. And there is the distinction between soul and mind:
>
> ... souls in general are the living mirrors or images of the universe of creatures, but minds or spirits are in addition images of the Divinity itself.

Leibnitz solves the mind/body problem by bringing God into this theory, thereby supporting the remark made by Lagrange as to how useful God can be. He is as dismissive of Spinoza as he is of Descartes, but he does see some similarity:

> [Spinoza says] 'No substance, not even corporeal substance, is divisible.' This statement is not surprising according to his system, since he admits but one substance; but it is equally true in mine, although I admit innumerable substances, for, in my system, all are indivisible or monads.

For Spinoza, mind and matter exist as different aspects of the same substance, which is the all-pervasive, immanent God. In the monadology of Leibnitz, Mind has become primary as the divine spark of the transcendent God. The British empiricists are remarkable in relegating mind to either a passive role with no God, or a hugely important role with God as an essential component.

In his *Essay on the Human Understanding*, John Locke (1632-1704) set out the empiricist position succinctly; 'No man's knowledge here can go beyond his experience.'

It is the experience of the external world that Locke is interested in and, here, the mind presents a problem. He sees the mind as a purely passive receiver of sensations but, somehow, out of sensations, ideas such as of time and space and of causation arise. Locke famously distinguishes between the primary and secondary properties of matter. Among the primary properties are solidity, extension, shape and mobility, whereas the secondary qualities are the immediate sensations of colour, sound, taste, etc. The latter, being subjective, can be ignored; only the primary qualities are objective enough to be the foundation of science. But those sensations of colour, taste, etc., are exactly those that the mind experiences directly. By dismissing these as irrelevant, Locke is already on his way to the description of a purely mechanical world that excludes any account of mental phenomenon.

But is there truly the distinction between primary and secondary qualities that Locke claims to be important? The Irishman, George Berkeley (1685-1753), aristocrat and Bishop of Cloyne, thought not. The idea of extension, for example, was entirely acquired from the senses of sight and touch. Each of Locke's primary qualities was, argued Berkeley, nothing more than an abstraction and not to be elevated to a universal principle. The extension of any object was just what was perceived and nothing more. According to Berkeley, the external world was, in fact, the very world that was presented to our senses, and since the mind was actively involved in what was perceived, the world itself depended for its actuality on being perceived. In his *Principles of Human Knowledge* (1710) Berkeley presented the startling argument that without the perceiving mind the world would not exist. 'All the choir of heaven and furniture of the earth—in a word, all those bodies which compose the mighty frame of the world—have not any substance without a mind.'

Matter as substance had no independent existence. The Latin sums it up neatly—*Esse est percipi,* To be is to be perceived. But what happens to things when nobody is looking? Do they continue to exist? The verse in the *Book of Limericks* gets hold of the problem neatly.

> There was once a man who said 'God,
> Must think it exceedingly odd
> If he finds that this tree,
> Continues to be,
> When there's no one about in the quad.'

Berkeley solves this problem by introducing the ever-perceiving God, as the limerick explains:

> Dear Sir, Your astonishment's odd:
> I am always about in the quad.
> And that's why the tree
> Will continue to be,
> Since observed by Yours faithfully, God.

This revolutionary solution to the problem of mind and matter was never likely to appeal to the Anglo-Saxon mind-cast. As Byron put it:

> When Bishop Berkeley said 'There was no matter',
> And proved it—'twas no matter what he said.

Talking of Bishop Berkeley's theory of the non-existence of matter, Boswell observed that though they were satisfied it was not true, they were unable to refute it. Johnson struck his foot against a large stone, till he rebounded from it, saying; '*I refute it* thus.'

Come the eighteenth century, it was high time that God ceased to figure in any account of what the New Science was about. With God out of the way, David Hume (1711-76) could reveal what empiricism was really about. He thought that Berkeley had done good job in dismissing Locke's distinction between primary and secondary qualities. He also appreciated Berkeley's argument that universal ideas are nothing more than ideas of particular things. In two books — *A Treatise on Human Nature* (1739) and *Enquiries concerning Human Understanding* (1748) — Hume managed coolly to demolish the very foundations of the New Science. All experience consists in being a succession of sense impressions, out of which the external world has to be constructed. This is to be done by perceiving relations between impressions that give rise to ideas of identity, quantity, space and time, causality and so on. But these ideas are not abstract universals, but are merely ideas of particular things and events. All ideas are derived from impressions and they are nothing but copies of them. 'When we entertain, therefore, any suspicion that a philosophic term is employed without any meaning or idea (as is but too frequent), we need to enquire, from what impression is that supposed idea derived?'

This is a powerful weapon to expose excessive, or pseudo-scientific claims. It certainly appealed to the logical positivists of the twentieth century, where it was successful in continuing the invisibility of the mind under the aegis of behaviourism.

Hume reduced the mind to be a receptacle for impressions. But he also talks about reason. The question arises, from whence comes reason out of a succession of sense impressions? It is not out there, it is an activity of the mind. To be sure, the ideas generated from impressions are pale features of the mind compared with the impressions themselves, but, nevertheless, they are there. How pale they can be can be illustrated by the idea of cause and effect. A's behaviour is observed to follow B in all experiments hitherto. The New Science says that therefore A's behaviour is caused by B. Hume objects to the 'therefore'. There are only impressions. The idea of causality can only be opinion. To suppose that there is something necessitating A's behaviour is a belief and nothing more. All of the ideas of the New Science are like this—mere opinion without certainty. In this sense, Humean scepticism destroyed all that the New Science had created.

One consequence of this was that a certain philosopher in Königsberg, one Immanuel Kant (1724–1804), was, as he put it 'aroused from his dogmatic slumber' and went on to create an unrivalled philosophic masterpiece in his *Critique of Pure Reason* (1781). Here, he agreed with the empiricist dogma that none of our knowledge can transcend experience, but he could not agree with Hume that, somehow, impressions caused ideas. Kant was sure that some knowledge was independent of experience, that it was *a priori*. Such knowledge, he claimed, had the characteristic features of necessity and universality. Without this sort of knowledge nothing certain can be said about the world, and we are reduced to Humean scepticism. Examples of this *a priori* knowledge are our intuition of space and time, of substance and, indeed, of mathematics and logic. 'For whence could our experience itself acquire certainty, if all the rules on which it depends were therefore empirical, and consequently *fortuitous*.'

(Yet another example of *ex nihilo nihil fit*.)

Kant rescued the mind from the passive role allotted to it by Locke and Hume. It becomes an active player, exploiting its *a priori* ideas to construct a science of the world. Yet the mind should not be regarded as some sort of pure, immaterial substance as some theology would have it. The mind is a human mind. Being finite it can have no clear notion of the infinite mind of God. All so-called proofs of the existence of God are false, resting as they do on the extrapolation of ideas that are valid only in our finite observable world. But in his *Critique of Practical Reason* (1786) he argues that God is necessary and that there must be a future life, for how otherwise could there be a just moral law whereby virtue was rewarded? In his ethical system he had two sorts of moral action or imperatives, hypothetical and categorical. The hypothetical imperative was to act to achieve a particular end. The categorical imperative was to act as an objective necessity, irrespective of end result. Kant needed God for his ethics, but not for his science.

Kant's philosophy is too dense and comprehensive for us, here, to do other, than as we have done, merely note a few highlights relevant to the mind/matter problem. Kant was well aware of the limitations of science — appearance will show us the existence of an object, but the thing-in-itself is forever beyond our knowledge. In his *Prolegomena* (1783) he points out that reason, too, has its boundaries.

> The enlargement of insight in mathematics and the possibility of new inventions extends to infinity; equally the discovery of new properties of nature, new forces and laws, by continued experience and unification of it by reason. But none the less we must not fail to see limits here, for mathematics only bears on appearances, and what cannot be an object of sensible intuition, such as the concepts of metaphysics and morals, lie outside its sphere ... Natural science will never discover to us the inside of things i.e. that which is not appearances; but natural science does not need this for its physical explanations.

Chapter Seven

Panpsychism

After the deep cerebrations of the famous philosophers of the Enlightenment we are no nearer solving the problem of mind and matter without dumping the whole problem in the lap of God. This may, indeed, be the only solution, but it offends against the Baconian injunction on that topic and, no doubt, most scientists of the nineteenth century and later would agree that there must be a better, or at least, a more interesting, way forward.

One of the first scientists to tackle the problem was a remarkable professor of physics — experimentalist, psychologist, writer and humourist — who originated the idea of psychophysics and advocated the doctrine of panpsychism. Well known at the time as a writer under his pseudonym, Dr Mise, he was, in fact, Gustav Theodor Fechner (1801–87), Professor of Physics at the University of Leipzig. In scientific circles he is best known for his application of mathematics to psychology via the Weber-Fechner Law. Weber noted that the sensation of touch varied in proportion to the logarithm of the stimulus, and Fechner showed that it worked for sight and sound. The law can be illustrated by the following musical example, familiar to all composers. Doubling the volume of perceived sound from two violins means increasing the number of violins, not to four, but to eight. The law is one of diminishing returns. Adding two violins to the original two doubles the volume of sound, but that is not how it is perceived. Fechner was also intrigued by his discovery that a stimulus must exceed a threshold for there to be any conscious sensation. In his *Elements of Psychophysics* (1860) he deduced the existence of a subconscious regime that received the sub-threshold stimulus.

The principle of his psychophysics was: to every psychic occurrence there corresponds a physical occurrence in the material world. Thus, matter and mind (or soul, Fechner makes little distinction) are inseparable, but the one must be conceived from without, the other from within. Here is an early example of the essence of the Principle of Complementarity. Organisms, viewed from without, were thoroughly physical, but viewed from within, they were souls. Thus, Fechner repudiated the transcendentalism of Kant in his dogma of the radically inconceivable essence of the thing-in-itself. Fechner claimed that the thing-in-itself could be known from within. In what he calls the *Daylight View* (1873), the entire material universe, instead of being lifeless, is inwardly alive and consciously animated. This doctrine, pure panpsychism, is to be opposed to the *Night View*, pure materialism. In his *Zend-Avista* (1851) he exhibits Hermetic fervour in his claim that the Earth and all the planets, the Sun and the stars, are all animate beings, in effect, the angels, the principalities and the powers of the celestial hierarchy.

Fechner, like Spinoza, sees mind and matter as two aspects of the same thing, but without Spinoza's pantheism. Fechner's analogy of his view from within and view from without is of someone inside a sphere who observed himself to be surrounded by a convex surface, while someone outside of the sphere sees a concave surface; but it is still the same sphere. This analogy may be fine for each of us humans, but not particularly useful to us for animals, plants and brute matter, where the only view available is the view from outside. In *The Soul Life of Plants* (1848) Fechner describes how plants respond to stimuli, how sometimes they behave irritably, how they exhibit life movement, all showing that that a plant has a soul. His defence against the criticism that plants do not have a nervous system and that, therefore, they cannot have any sensibility, is to make the point that music is not made only by stringed instruments, but also by hollow tubes, solid blocks and by drums.

(Modern biology is far from being dismissive of Fechner's intuitions about plants. Here are the titles of two recent scientific papers: 'Plant Neurobiology: An Integrated View of Plant Signalling' [Brenner *et al.*, 2006] and 'Kin Recognition in an Annual Plant' [Dudley & File, 2007]).

Fechner's conviction of the immortality of the soul is to be found in his still popular book *The Little Book of Life after Death* (1836). The sort of immortality he envisages has a curious double aspect about it, one aspect uncontroversial, the other more contentious. Both aspects emphasise the rather orthodox view that the quality of one's life on earth will affect the life of the soul in the great spiritual world. The first aspect is nothing more than the immortal reputation of outstanding actions. Examples are Shakespeare's creations and Napoleon's campaigns, awareness of these will never die. Therefore, Fechner exhorts the living to produce or do something memorable, as this will guarantee a kind of immortality. Nothing controversial about this, but whatever the achievement, all souls will survive. But if matter and mind are but aspects of an underlying reality, then if the body suffers dissolution, surely the soul will too. In *Zend-Avesta* Fechner answers this criticism with another analogy that relates the soul to the sound of a violin.

> You think, if a violin which has just been played upon, is broken up, then it is all over with the music: it dies away, never to sound again, and so dies away the self-conscious music of the human brain, when death destroys the instrument. But at the destruction of the violin, as also in the death of the man, there is something that you neglect, in looking only at that which is most obvious. The notes of the violin resound in the wide air, and not only the last note of

the music, but the whole of it. Now you suppose that, when the sound has gone by you, it has died away; but anyone standing at a greater distance can still hear it, therefore it must still exist; one who stands too far away will not hear it at all, but not because it has ceased to be; the sound merely spreads itself out too widely, becomes too feeble to be heard at a single spot; but imagine your ear accompanies the sound and spreads itself out with the widening circle of vibration, then you would continue to hear it. It is never extinguished; it remains forever. The narrowly bounded violin has spread its music to infinity.

The quantum physicist will recognize an analogy here with Schrödinger's wavefunction which never collapses but retains its coherence forever. If Fechner had been born a century later he might have replaced his violin with the wavefunction of some submicroscopic particle.

If we believe in the soul, Fechner tells us in *The Three Motives and Grounds of Faith* (1863), we do so because of one of three motives:

> The historical motive: we believe what we have been told, what was believed before us, and now is believed all around us.
>
> The practical motive: we believe what we are pleased to believe, what is serviceable to us and profitable.
>
> The theoretical motive: we believe what we find grounds in experience and reason for believing.

(Actually, we may believe because of a mixture of all three.) The historical motive is familiar, if somewhat mindless; the practical motive could turn into an art form; the theoretical motive accounts for our believing in science.

Science cannot do other than to treat its sphere of research in terms of out-and-out materialism. God and the soul are firmly banished and play no part in the scientific method.

Matter obeys the Laws of Physics, which means that its motion, if not as deterministic as was once thought, still is regulated by the Laws of the Conservation of Momentum and Energy. But as the physicist Ernst Mach (1838–1916) needed to say, in his book *Science of Mechanics*, 'The science of mechanics does not comprise the foundations, no, nor even a part of the world, but only an aspect of it.'

Nevertheless, many scientists, and many who see how science works, are solidly convinced of materialism, the doctrine that everything that exists is matter and motion. In which case how is the phenomenon of consciousness to be explained? One of the illustrious founders of quantum mechanics, Erwin Schrödinger (1887–1961) describes realistically in his book *Mind and Matter* (1958) the response of the materialist to the question of the material process that is directly associated with consciousness.

> A rationalist may be inclined to deal curtly with this question, roughly as follows. From our own experience, and as regards the higher animals from analogy, consciousness is linked up with certain kinds of events in organized, living, matter, namely, with certain nervous functions. How far back, or 'down', in the animal kingdom there is still some sort of consciousness, and what it may be like in its early stages, are gratuitous speculations, questions that cannot be answered and which ought to be left to idle dreamers. It is still more gratuitous to indulge in thoughts about whether perhaps other events as well, events in inorganic matter, let alone all material events, are in some way or other associated with consciousness. All this is pure fantasy, as irrefutable as it is unprovable, and of no value for knowledge.

But he goes on:

> After Spinoza the genius of Gustav Theodor Fechner did not shy at attributing a soul to a plant, to the earth as a celestial body, to the planetary system etc. I do not fall in with these fantasies, yet I should not

like to have to pass judgement as to who has come nearer to the deeper truth, Fechner or the bankrupts of rationalism.

Whatever one may think of Fechner's analogies, it is surely evident that the question of how the mind and the body interact must be of vital concern to any thinking person. The eminent psychologist, William McDougall (1871–1938) puts it like this in his book *Body and Mind* (1911):

> If a man is to live, he must act: and if he must act, he must govern his actions in accordance with conceptions of his own nature and of the world in which he is set, concepts of whose validity he can have no absolute guarantee, and which he must choose, develop, reshape or reject, according as he finds them more or less efficient guides to successful action. And of all conceptions, the conceptions of the nature of, and of the relations between, mind and body are those which in the long run affect most profoundly, and are of the first importance for, this guidance of conduct; for they must always exert a determining influence upon man's view of his place in the world.

He goes on to explore the mind/body relationship in some detail and to frame a spirited defence of animism, the doctrine that soul and body reciprocally influence one another.

One answer to the question of the mind/body interaction suggested by the materialist is the doctrine of parallelism: psychical and physical processes run parallel to each other. This is like the view of Leibnitz without the God-given pre-established harmony. The most radical doctrine is that of epiphenomenalism. Somehow, in complex structures like the brain, the motion of matter in the brain gives rise to consciousness, but consciousness, being always preceded by brain activity cannot have any effect on the functioning of the brain. According to this doctrine, a complex enough computer could be conscious, though how could you tell? If evolution is true, the development of an increasingly complex conscious mind that was merely epiphenomenal makes no sense if it could not act in the battle for survival. McDougall is not alone among philosophers in rejecting epiphenomenalism.

Reductionism, so successful elsewhere in science, suggests a kind of atomic hypotheses to explain parallelism. It consists of regarding consciousness as a composite of sensation, idea and feeling. For example, the observation of an object might include the sensation of colour, the idea of size and why it was there, and perhaps a feeling of how annoying that was. Each 'atom of consciousness' could then be related to a particular element of brain activity. McDougal points to the evidence of the existence of sub-brains to refute this idea. One might suppose that the visual signals from both eyes would be physically fused into a single input in the brain, but this is not the case; what fusion occurs does so in the psychic sphere. Let a light that flickers with increasing frequency be observed. Beyond a certain threshold frequency, say sixty per second, the sensation of flickering disappears. Now suppose it is arranged that alternate flashes are directed at the right and left eyes, so that each eye receives thirty per second. Were the signals processed at one site, the site would receive sixty flashes per second and there would be consequently, as before, no flickering. But this is not the case — flickering is observed. Thus, it is clear that each eye is connected to its sub-brain and its input processed singly before psychically being combined into one.

Indeed, there are overwhelmingly many experiments that show that optical perception is not as simple as the reductionist picture would suggest. Think of the experience of observing complex random-dot stereograms through spectacles with a green filter for one eye and a red filter for the other. It takes time to see the pattern that gradually emerges. The visual input to each eye is constant, so something goes on in the brain, independent of the basic stimulation, before there is perception. And then there are patterns that are deliberately ambiguous, such as a picture of a staircase which can appear as viewed from underneath or from above, as a matter of choice. A famous example is Wittgenstein's duck-rabbit sketch; viewed one way and the rabbit's ears become the beak of a duck, viewed the other way, the beak becomes the ears of a rabbit. Which one sees is a matter of volition, the mind deciding what pattern of physical activity occurs in the brain. Wittgenstein points out that a mental image never has that ambiguity. All of which supports an interactionist model of mind and brain but that the mind always appears as a unity.

What William James (1842–1910) objected to in Cartesian duality, among other things, was the idea of mind as a substance. James was convinced that there had to be a naturalistic, non-immaterial, understanding of the mind-body problem, so questions about the existence of the soul and where in the brain it might be situated were not useful. He saw the mind and its activities as having attributes more of becoming than being, an intuition that crystallized into the concept of 'a stream of consciousness', a phrase that has gone seamlessly into the language, and an idea that has proved influential in promoting a more realistic description of psychic phenomena. Although he was sympathetic to Fechner's ideas, Darwin's ideas of evolution convinced him that mind had evolved along with the body, so a naturalistic (materialistic) theory that included the survival advantages of enhancing the power of mind was preferable. He was unhappy about Descarte's soul-body duality, but he was equally unhappy about monism, seeing Spinoza's variant as absolutist, deterministic and timeless. In contrast James advocated a distinctly different vision in his book *The Pluralistic Universe* (1909) which he called radical empiricism.

> I may contrast the monistic and pluralistic forms in question as the 'all-form' and the 'each form'. The 'each form' is our human form of experiencing the world and it makes God only one of the 'eaches', and more approachable as a consequence.

Panpsychism

The existence of the soul is another mystery altogether. Those who argue that we have a soul point to the fact that we appear to ourselves as such a unity. An argument due to a defender of animism, R. H. Lotze, and quoted by McDougall, is more trenchant. 'Our belief in the soul's unity rests not on our appearing to ourselves as such a unity, but on our being able to appear to ourselves at all.'

Yet, as Charles Sherrington (1857–1952) observes in his book *Man on his Nature* (1940) there are severe problems in having a clear conception of soul.

> We are made up of billions of cells, each a living entity, aware of its surroundings, needful of food and energy. Yet they die and are replaced. When we die they do not—at least immediately—the beard continues to grow etc. Organs, removed from the body and surrounded by nutrient fluid, continue to live. Where does the mind—the soul—come into all of this?

And cells are composed of molecules, which are composed of atoms, which are composed of protons, neutrons and electrons and the protons and neutrons are composed of quarks.

> [Which] have in themselves not the faintest elements of the visual—having, for example, nothing of 'distance', 'right-side-upness', nor 'vertical', nor 'horizontal', nor 'colour', nor 'brightness', nor 'shadow', nor 'roundness', nor 'squareness', nor 'contour', nor 'transparency', nor 'opacity', nor, 'near', nor 'far', nor visual anything—yet conjure up all of these.

Sherrington states the dilemma of the physiologist beautifully. 'It is the dilemma of us all.'

'Do molecules blindly run?' asks Alfred North Whitehead (1861–1947) in his book *Science and the Modern World* (1926). Whitehead was extremely aware of the shortcomings of the prevailing materialism of modern science, and looked forward to a description of the world in terms of organism, the distinguishing feature of the world, like organs in a living being, functioning according to the pattern of a powerful structure. Following this idea he was lead to the remarkable possibility of organism affecting the behaviour of electrons.

> In the case of an animal, the mental states enter into the plan of the total organism and thus modify the plans of the successive subordinate organisms until the ultimate smallest organisms, such as the electron, are reached. Thus an electron within a living body is different from an electron outside it, by reason of the plan of the body.

Whitehead, mathematician and one of the last of the Cambridge Platonists, had more than a touch of the Hermetic philosophy, in particular its dogma of macrocosm-microscosm — within each of us is the whole universe.

> In a certain sense, everything is everywhere at all times. For every location involves an aspect of itself in every other location. Thus every spatio-temporal standpoint mirrors the world.

> If you try to imagine this doctrine in terms of our conventional views of space and time, which presuppose simple location, it is a great paradox. But if you think of it in terms of naïve experience, it is a mere transcript of the obvious facts. You are in a certain place perceiving things. Your perception takes place where you are, and is entirely dependent on how your body is functioning. But this functioning of the body in one place, exhibits for your cognisance an aspect of the distant environment, fading away into the general knowledge that there are things beyond. If this cognisance conveys knowledge of a transcendent world, it must be because the event which is the bodily life unifies in itself aspects of the universe.

Platonic transcendentalism apart, Whitehead's intuition of a less mechanical and a more organic science of the universe finds at least a kind of political support from J. S. Haldane. In his Presidential Address in 1908 to the Physiological Section of The British Association of Science, he advocates a new scientific hierarchy. 'For Biology we must clearly and boldly claim a higher place that the purely physical sciences can claim in the hierarchy of the sciences—higher, because Biology is dealing with a deep aspect of reality.'

We are still a long way from even catching a glimpse of a unified physico-organic theory of the universe. If this implies some sort of panpsychism, the best that can be said is that panpsychism is not ruled out of hand by physiologists like Sherringham.

> There seems no clear lower limit to mind. The mind which we can infer and, so to say, observe without difficulty in our daily intercourse, is that of our human kin. To it our experience of ourselves is guide. Our motor behaviour and theirs are interpretable each to each. But security of inference regarding mind fades as traced downwards along the scale of being. Ultimately mind so traced seems to fade to no mind. It becomes so meagre that the problem becomes that of trying to prove a negative.
>
> Is the ability to learn an indication of mind? '... is mind recognizable in cartilaginous fishes? The reply comes "fish can learn". That inference perhaps allows them mind.'

The motor behaviour of single cells suggest that; 'Microscopic single-cell life, without sense organs and without nervous system, can learn.'

'Mind as attaching to an unicellular life would seem to me to be unrecognizable to observation: but I would not feel that permits me to affirm it is not there.'

Even if it is not possible to rule out in primitive life forms, mind seems to become an increasingly useless piece of baggage to carry around, except for the thought that if mind is totally absent in the lowest life forms, there arises the question, 'In which higher life form does it first appear?'

Chapter Eight

The Mind-Body Interaction

The commonsense view of all this seems to me to be that mind and matter are real distinct entities, that interact in spite of their wholly disparate qualities, and that that interaction is a mystery, one that has remained so in spite of the thousand arguments of centuries of scientists and philosophers. A reformed science would accept the existence of mind as obvious. As for the immaterial soul, its existence or non-existence, like the existence or non-existence of God, is simply a matter of belief.

Belief comes in all sorts of forms, as Fechner reminds us. There are beliefs that we grow up with, beliefs shared by those around us, some of them in our very bones that remain with us forever, others that fail to survive the development of a critical mind. An example of the first type is the belief in the probability of serious injury occurring if one steps off a cliff; an example of the second may be the belief in the existence of a loving God. In maturity we acquire beliefs in the laws of science, many of which get into our bone-marrow as easily as falling off a cliff—belief in the simple rules of billiard-ball collisions, or belief in the thermodynamic arrow of time. Others, especially those that claim timeless universality such as the laws of electromagnetism, the law of gravity, the law of natural selection, we believe more tentatively. These are beliefs that can be modified by reason and experience. We believe that some actions of man are good and others are bad, and these beliefs are much less open to being modified by reason. There is even less scope for a role for reason in aesthetic judgement: 'You should like this painting, this piece of music, this poem, because ... ' Well, this has never worked for me.

Reason and intelligence are qualities of the mind that are accompanied by intuition. Where intelligence rules, intuition has no place: where intuition is the creative force, intelligence is irrelevant. In a word, intelligence and intuition are complementary qualities of the mind. As has been pointed out before, mind and matter also stand to one another as complementary entities. It would seem that the Principle of Complementarity, discovered in quantum mechanics in relation to the dynamic quantities momentum and position on the one hand and energy and duration on the other, has a significance in the world far beyond quantum theory. It implies that a full understanding of mind and matter, reason and intuition, cannot be obtained without recognizing the fundamental complementary nature of the elements involved. A complete description of the neurology of the brain will say nothing about consciousness; feelings, beliefs and thoughts cannot be understood in terms of the firing of neurons; yet both consciousness and the electro-chemical activity of the brain are Spinoza-like aspects of the same entity, neither mind nor matter but both. In the same way, rational thinking is deduction from intuitively conceived axioms of belief; the two elements of creative thought are each essential, but they are complementary. The same is true of being and becoming. Think of a melody. It has its being as a name and as a musical score, but also a becoming as it is played or sung. The philosopher Henri Bergson has been eloquent on how science focuses on being, but has difficulties with becoming, how it has clear ideas about time reduced to a space-like coordinate, but not of duration. Being and becoming, time and duration, are governed by the Principle of Complementarity, one driving out the other, yet both necessary for a comprehensive understanding of the world.

The world appears to present itself to us as a plurality of dualities where, in each duality, the elements are in a complementary relation that defies all attempts to reduce the duality to a unity. It would seem that our understanding has to be in accord with a General Principle of Complementarity which reads:

> Our most general and fundamental knowledge of the world must be expressed in terms of complementary pairs of concepts, each of which can be better defined only at the expense of a loss in the degree of definition of the other.

It implies that each basic entity of which the world consists never presents to us a single aspect. In the case of the wave-particle duality we call the basic entity the electron; in the case of the mind-brain dualism we may conveniently call the basic entity the soul, one aspect of which is the immaterial form of Plato, the other, the bodily form of Aristotle. Thus do Plato and Aristotle metaphorically embody the Principle! And as the electron has other attributes — electric charge and magnetic moment — so may the soul.

This elevation (if that is the right word) of a Principle, that owes its existence to the empirical discovery of the wave-particle duality, to a nostrum applicable to certain metaphysical concepts is unlikely to be regarded as immediately compelling. It has the aura of a Fechner-like analogy, but none the worse for that. Its practical virtue as regards the mind-body problem is to provide an argument for regarding all attempts at explaining how mind and body interact as pointless, since mind and body are the merely complementary attributes of the same thing.

Any attempt of this sort would be reminiscent of the suggestion made by Louis de Broglie for resolving the wave-particle problem. His idea, followed up and elaborated by David Bohm (1917–92), was that both wave and particle existed as separate entities (dualism) and that the wave acted as a kind of force-field that determined the motion of the particle. This picture of the pilot wave went rapidly out of favour as quantum theory developed, but the idea has never died. Indeed, it has been elaborated into the so-called 'Causal Interpretation' of quantum mechanics, in which uncertainty of prediction, something about quantum theory that Einstein never liked ('God does not play dice!'), is accounted for by the uncertainty in the starting conditions. In this it does not differ from the classical physics of certain non-linear systems where a tiny change in the initial dynamics can produce an enormous difference in the result. In these non-linear systems, classical determinism simply does not exist in practice. Quantum theory, causal or acausal, shows that strict determinism exists nowhere in the physical world.

In the de Broglie-Bohm causal interpretation, the wave has become the active principle, the particle taking a passive role. A similar dichotomy has been proposed by Karl Popper and John Eccles to solve the mind-body interaction. In their book *The Self and Its Brain* (1977) they present a dualist view of a self-conscious mind interacting with a fully material brain, with the self-conscious mind calling the shots. The mind is distinguished by its unitary character which it maintains. Its experiences have a relationship with the neural events in a special structure of the brain which they call the liaison brain, but are not identical with them.

> It is proposed that the self-conscious mind is actively engaged in searching the brain events that are of its present interest, the operation of attention, but it is also the integrating agency, building the unity of conscious experience from all the diversity of brain events. Even more importantly, it is given the role of actively modifying the brain events according to its interest or desire, and the scanning operation by which it searches can be envisaged as having an active role in selection.

This proposal, though formulated in modern language, is surely nothing less than an argument for animism, with the interaction with the governing mind and the passive matter proceeding via the immensely active liaison areas of the brain. They claim that their hypothesis is science because it is based on empirical data and is objectively testable. The implication is that there are laws of the mind-brain interaction waiting to be discovered, and they predict that the dominant role of the self-conscious mind will not be falsified.

Ever since the French neurosurgeon Broca (1824–80) discovered, in the mid-nineteenth century, that the left-hand side of the brain was associated with speech, the mapping of the specialized areas of the brain, those that deal with vision, with hearing, with touch, etc., has gone on apace. In recent times physics and electronics have come to the aid of neurophysiology in the invention of Positron Emission Topography (PET), which displays the areas of the brain that become active under various intellectual stimuli. This technique is revealing impressive information of the brain's activity but, so far, not offering insights into how that activity translates into self-awareness, memory, will, emotion, imagination, and all the manifold attributes of the mind. The task, if remotely possible, is in any case complicated by the ability of the brain to change and to adapt to new experience. My brain is not the same brain today as it was yesterday, but somehow I am still me. Discovering the laws of the mind-body interaction, if that is a coherent concept as Popper and Eccles suppose, is far from being realized.

Most advocates of materialism would agree that, indeed, there are laws relating to the mind-brain interaction, though they naturally reject the dualism. An exception is the philosopher Donald Davidson (1917–2003), who is happy with the materialist claim that mental events are identical to physical events, and that only physical events exist. However, he points out an important difference between understanding mental and physical events. Any psychological interpretation of belief or intention cannot just accept the particular belief of intention on its own. If the interpretation is to be meaningful, account must be taken of the coherence of these events with other beliefs and intentions of the person. Mental events cohere to varying degrees with other mental events, otherwise there would be a sense of randomness about consciousness which there is not. Thus there is a holistic normative constraint placed on psychological explanation that is entirely absent in physics. Therefore there can be no laws of the mind like physics, and therefore there can be no strict laws that can be formulated of the psycho-physical interaction. Insofar as Davidson espouses materialism, his approach is monistic; insofar as his approach denies the possibility of psycho-physical laws, it is anomalous monism.

Davidson's materialism is considerably subtler than most variants of materialism. The bog-standard sort is reductive materialism or *physicalism*, in which everything is reducible to physics and every psychic property is equivalent to or identical with a conjunction of physical properties. An extreme form is Paul Churchland's *eliminative materialism* in which he forecasts that what he calls the folk-language of psychological discourse—talk about belief, intention, feeling, pain—will ultimately become redundant as the study of neurology of the brain advances, and can be replaced by a description of firing neurons. This extreme but perfectly logical deduction of physicalism has to be judged against the attempt by Frank Jackson to demolish physicalism once and for all. He regales us with the following story.

> Mary is confined to a black-and-white room, is educated through black-and-white books and through lectures relayed on black-and-white television. In this way she learns everything there is to know about the physical nature of the world. She knows all the physical facts about us and our environment in a wide sense of 'physical' which includes everything in completed physics, chemistry and neurophysiology; and all there is to know about the causal and relational facts consequent on all this,

> including, of course, functional roles. If physicalism is true, she knows all there is to know. For to suppose otherwise is to suppose there is more to know than every physical fact, and that is just what physicalism denies.
>
> It seems, however, that Mary does not know all there is to know. For when she is let out of the black-and-white room or given a colour television, she will learn what it is like to see something red, say.

And in his article in *Contemporary Materialism* (1995, ed. Moser and Trout) he goes on to rebut the criticisms of Paul Churchland. Game, set and match, you might think, but if you do, Churchland doesn't. The ball is still being swiped across the net. But then, what philosophic game ever finishes?

One can abandon strict reductionism without abandoning materialism by adopting the idea of *supervenience*, the view that nothing mental can happen at all unless something happens at the physical level. Sensations are associated with particular areas of the brain, and these, in turn, can affect the mind. If brain events cause mental events, this is cause-and-effect, something that is deeply part of physics, in which energy and momentum must be conserved. Thus, whatever the details, the mind-body interaction, if not crudely reductive, yet must be essentially physical. But psychic properties differ from material properties in at least two important ways, the holistic nature of mental events to which Davidson drew our attention, and the special causal or functional rules that pertain to mental events.

Hilary Putnam puts the *functionalist* case succinctly: it is not the stuff of which it is made that makes a mental process the thing it is, rather what it does, what sort of cause-and-effect it has. So forget about the firing of neurons (always there, of course, in the background) but concentrate on function. So far, this does not rule out the immateriality of the mind, though Putnam would use Occam's razor to favour materialism. If so, and if function is the essence, one can conceive of mind existing without a brain, a thought that naturally delights some computer scientists and students of Artificial Intelligence. Functionalism implies multiple realizability for the appearance of mind — maybe cars, barometers, kettles — anything that supports a function. We have here a vision of functional panpsychism! More soberly, we may see mind as being analogous to a computer programme — something not reducible to physical and chemical events, mind as the software of the brain. This is materialism perhaps at its most convincing.

But there are deep problems in seeing mind as a computer programme. The latter is merely a set of instructions, an algorithm, beyond which the computer can do nothing. This limitation reminds us of the conviction of Lady Lovelace in her notes on Babbage's machine: such a machine is entirely mechanical (literally so, in the case of Babbage's engine) and to speak of such a thing as having intelligence is nonsense. Yet Alan Turing, famous *inter alia* for his role in cracking German codes at Bletchley Park, could envisage a behaviourist experiment whereby, separated by screens, a person communicated electronically with an advanced computer. If the person could not distinguish whether an actual person or a computer was on the other side of the screen then, surely, the computer could be described as intelligent.

A famous argument against this that became known as the *Chinese Room* was advanced by John Searle. Imagine a person, knowing no Chinese, isolated in a room but equipped with a set of rules in English for manipulating Chinese symbols that tell a story to him posted through a slot. Without knowing a word of Chinese, but following all the rules of manipulation, he could correctly answer questions put to him about the story, even though he would not have a clue of what the story was about. Like a machine following instructions, he would not have the least understanding.

Another argument against the idea that minds were machines was advanced by J. R. Lucas, who claimed the

Gödel's theorem proved the point. Kurt Gödel startled the mathematical world in 1931 with his theorem that in any consistent system which is strong enough to produce simple arithmetic there are formulae which cannot be proved in the system, but we can see that these formulae are true. Within the system we can always discover a formula 'G' that says 'G' is unprovable. If the formula turns out to be provable within the system, we have a contradiction and the system is inconsistent. If, on the other hand, 'G' is unprovable within the system, the system is incomplete. Adding a new axiom so that this particular 'G' is now provable within the system can only be a temporary expedient, for there will always be a 'G' that is unprovable. Lucas's point was that no computer algorithm, which uses arithmetic, would be complete; there would be truths beyond the ken of any machine which, nevertheless, could be known by a mind. It is an argument that holds good only if the mind is seen as capable of arithmetic consistency and with no limit to its insight. Critics of Lucas's claim doubt whether this is true of the mind, and particularly whether the minds of mathematicians are infinitely creative.

But, thinking of Mary coming out of her black-and-white room and seeing red, one might wonder what it is like for a computer to be something that can see red. The question 'What is it like to be ... ?' gets to the nub of the difference between mind and matter. Thomas Nagel asks 'What is it like to be a bat?' Impossible to say. No amount of bat neurology will explain what it is like for a bat to navigate the world using ultrasonics. Or, for that matter, no amount of neurology will explain what it is like to be me. Only I know. That's a fact of the world. All those elements of conscious experience directly accessible only from the first-person point of view, the qualia (singular quale), are real. In his book *The View from Nowhere* (1986), Nagel contrasts qualia as the view from the first-person with the view from nowhere, the objective view, somewhat as Fechner juxtaposed the Daylight view with the Night View.

Panpsychism, the doctrine that Fechner so eloquently advanced, has its support in modern philosophy apart from Nagel. Galen Strawson contends that the adoption of physicalism actually implies an acceptance of panpsychism. His arguments and those of his critics can be found in *Consciousness and its Place in Nature*. Certainly, thought, as Sir Arthur Eddington remarks in his stimulating book *The Nature of the Physical World*, is one of the indisputable facts of the world. He would claim:

> ... that those who in the search for truth start from consciousness as a seat of self-knowledge with interests and responsibilities not confined to the material plane, are just as much facing the hard facts of experience as those who start from consciousness as a device for reading the indications of spectrometers and micrometers.

Critics, however, worry about what it means for a quark or a photon to be conscious. As well they might! But quarks and photons are creatures of theory. Photons are there to explain pointer readings; quarks are not even that. Quarks are there to explain the pattern of elementary particles and cannot, by their very hypothetical nature, be 'seen' outside of the particles of which they are the constituents. To suppose consciousness in the fundamental entities of physics is scarcely different from supposing consciousness in pi, or the fine-structure constant. But physics tells us that everything in our everyday experience is made of quarks, gluons, photons and electrons. Where does the psychic element come in? Only in the organization that is life?

At least we know what consciousness is. Each of us directly experiences consciousness, and we know that it is very different from known physical processes with their mechanistic character. James has called it a stream, rather than a sequence of events, a view that is eloquently shared by Henri Bergson (1859–1941), who was at pains to distinguish *duration,* experienced in consciousness, from the concept of time used in mathematical physics. In the former, time flows seamlessly, in the latter it associated with this-then-this-then-this ... Bergson has so many interesting things to say about consciousness that it is worthwhile to consider, however briefly, some of the aspects of his philosophy that are relevant to us here.

His unique account of reality begins with an analysis of time and space. Any rational concept of time and space must involve quantification which, inevitably, involves the idea of numbers. Bergson argues that the use of numbers implies seeing things as having a multiplicity of units, each similar to the other, yet distinct, and this is entirely consonant with our intuition of space. In the three-dimensional space of mechanics, the units can be as small as the description of the events demands, but they remain countable.

But what of time? Newton provides the description of the time used in classical physics in his *Principia* (1687): 'Absolute, true, and mathematical time, in itself, and from its own nature, flows equably without relation to anything external.'

And in his *Essay Concerning Human Understanding* (1690), John Locke states: 'Duration is but as it were the length of one straight line extended in infinitum.'

The analogy with 'line' gives the game away. Mathematical time is regarded as a dimension like space, and is quantified in the same way, that is to say, in terms of units. In mechanics, the equable flow of time is modelled digitally, in much the same way as it is simulated in the cinema with its rapid sequence of slides. If the flow of time is really like this, then the mathematical model is a good one. The idea that time is actually atomic, and not continuous, goes back to the Pre-Socratics, plausibly to the Pythagoreans. Zeno was at pains to refute this, pointing out that if it were true, then motion would be, indeed, impossible. Curiously, the atomists did not challenge Zeno's argument. But any concept of time that eliminated motion is too strange to take seriously. Nevertheless, mathematics has to adopt a provisional atomicity of time in order to quantify it. But, says Bergson, this is not the way each of us experiences duration. There is no 'now', then 'now', then 'now': the flow is experienced as truly continuous, a 'present' not as a being, but as a becoming. Yet he is fully aware that a rational concept of time entails digital quantification. Science, the product of man's intelligence, cannot do otherwise.

The difference between the experience of living and the knowledge of matter that science produces is unbridgeable. In science the state of matter finds its complete explanation in the state immediately before. Not so of the living body —all its past, its heredity, its long history—all have to be taken into account. In this, Bergson is at one with Davidson. But the difference between the experience of living and scientific knowledge goes much deeper. *There is no state immediately before*, there is only change and duration, like that directly experienced in consciousness. Matter continually interacts with matter, an interaction that is continuous throughout the universe. Looked at as an undivided whole in the universe, matter must be a flux rather than a thing. By its nature, physics can deal only with being, not becoming. Its differential calculus handles motion by defining velocity and acceleration at a given instant of time, in an eternal present. No matter how small its dt and dx are, the flow of real time, experienced by the conscious mind, is not captured. One consequence of this is that the sign of the space-like coordinate that serves for time in the differential equations of physics can be plus or minus without altering the laws of mechanics. Mathematical time can go backwards. In general relativity it can even go round in circles. This not the time that living things are aware of. Because of its inability to incorporate real time in its laws, physics, and science in general, cannot hope to describe the conscious living organism. Mechanical explanations of bodily matter are fine, but they can say nothing useful about the mind.

In a world of becoming, consciousness embodies choice.

> Everything seems, therefore, to happen as if consciousness sprang from the brain, and as if the details of conscious activity were modelled on that of cerebral activity. In reality, consciousness does not spring from the brain: but the brain and consciousness correspond because equally they measure, the one by the complexity of its structure and the other by the intensity of its awareness, the quantity of choice that the living being has at its disposal.

Bergson sees science as the natural activity of humans that helps them to get along in life and teaches them how to manipulate matter to their advantage. It is why intelligence has evolved. Its role is spelled out in *The Creative Mind* (1965 edition):

> Our intelligence is the prolongation of our senses. Before we speculate, we must live, and life demands that we make use of matter, either with our organs, which are natural tools, or with tools, properly so-called, which are artificial organs. Long before there was a philosophy and a science, the role of intelligence was already that of manufacturing instruments and guiding the action of our body on surrounding bodies. Science has pushed this labour of intelligence much further, but it has not changed its direction.

But intelligence can create only conceptual knowledge about things that are remote from the direct experience of our conscious life. What *really* exists is not things, but things in the making. Here, Bergson is pure Heraclitus rather than Plato. He sees the Forms of Plato and the Categories of Kant as setting the world in a static mould, whereas the world is a world forever becoming, forever creating itself. True reality can be known only through intuition, which means thinking in duration. We are living matter and can therefore *know* the living and the matter. There are echoes here of the Hermetic philosophy—the like can only be known by the like. The force of intelligence, through science, can explain matter;

only through a metaphysics involving intuition can an explanation of life be found. As William James puts it in *The Pluralistic Universe*: 'Direct acquaintance and conceptual knowledge are thus complementary of each other; each remedies the other's defects.'

Which recalls the 'knowledge by acquaintance and knowledge by description' of Bertrand Russell. We may remark that here is the Principle of Complementarity in action through the pairs, being and becoming, intuition and intelligence, matter and mind. Bergson is adamant that only through employing the mind's complementary powers on intuition and intelligence can the living organism be understood.

Bergson points out the difficulty of employing intuition as a metaphysics. Language itself is a language of being, not becoming. Instead of 'The child becomes the man', it would be more real to say 'There is becoming from the child to the man'. In this, 'becoming' becomes a subject. The trouble with intuition is that, although it undoubtedly exists, it is wordless. Nevertheless, Bergson has put his finger on a hugely significant point, and is unashamedly a dualist.

Mind and matter exist in the world, as unlike as anything could be this side of opposites. They are so unlike, one has to ask how is it possible that they interact. Yet they do, as we all know sometimes to our cost. The infelicity of that last irresistible scotch, and one knows about it the next morning; the unwise decision to play singles tennis with one's ever so much younger son, and the muscles know about it for days. Materialists, of course, have no choice but to search for a scientific explanation of that interaction, unless, like Davidson they despair of the task. Dualists, like Popper and Eccles, seeing two disparate things that interact, also must search for a scientific explanation, though, in their case, their intention seems incoherent, since dualism claims that mind and matter are utterly distinct. Materialism, despairing of finding an explanation, can always bury the problem in the sands of epiphenomenalism or even behaviourism; idealism can always bring in God. Materialism and idealism are the two poles towards one or the other the commonsense belief in dualism tends. Alternatives are offered by Spinoza and by Bergson.

The problem of the interaction can be avoided only if it is seen that mind and body are but aspects of a single thing. In this case much hangs on what is meant by 'aspect' and what is the nature of 'a single thing'. In Spinoza's pantheism, the single thing is the immanent God-Nature, mind and matter are but two of God's aspects presented to us. I have suggested that God may be replaced with a (secular) soul, whose bodily and mental attributes form a pair that obeys the Principle of Complementarity, so that, just as it is pointless to see the electron as either a wave or a particle, so it is pointless to see the individual person, the soul, as either mind or matter. Bergson sees a similar complementarity, but with a much more active role of consciousness, one that we will explore further when we consider origins.

Chapter Nine

A Strange New Science

At the beginning of the twentieth century there were two earth-moving events. One was the discovery that Newtonian mechanics, the bible of classical physics, needed to be modified in the light of Maxwell's electromagnetic theory, the other was that particles were waves and waves were particles. Classical physics began to have serious problems. There were two outstanding ones. The existence of atoms had, at long last, become accepted and the discovery of the electron and ions in gas discharge made it natural for physicists to speculate that the lightest, and presumably the simplest, atom would consist of an electron somehow bound to an equal and opposite electric charge, the proton. How could the atom be stable? Why didn't the electron and positive charge collapse together under their electrostatic attraction? Ernest Rutherford's experiments with alpha particles showed that the hydrogen atom was mainly empty space and the proton was concentrated at the centre. This merely exacerbated the problem. Why didn't the electron spiral into the proton and neutralise it? Even if the electron orbited the proton, like a planet round the sun, it would radiate energy according to well-established electromagnetic theory since it was being continuously accelerated towards the centre, and accelerated charges, as we know from alternating currents in an aerial, radiate electromagnetic waves. The stability of atoms could not be understood by classical physics, which meant that the existence of matter itself was a mystery. Not a very good situation!

The other problem was understanding the spectrum of so-called black-body radiation. At all temperatures above absolute zero any material object radiates electromagnetic radiation. If its temperature is the same as the temperature of its surroundings, it absorbs thermal radiation as much as it emits. Mathematically it is useful to imagine an ideal chunk of matter that absorbs all radiation equally well. Such is known as a black-body. If the body is at a temperature above its surroundings the properties of its emission can be distinguished from the background thermal radiation. What can be measured is the spectrum, that is, what wavelengths of light are emitted and the intensity — the energy per second per square metre — at each wavelength. It turns out that the higher the temperature the greater the intensity at all wavelengths, and the wavelength of the peak of the intensity shifts to shorter wavelengths; so that, in a fire, coals glow red and when they are very hot they glow yellow. The intensity falls away with wavelength on either side of the peak. The account that classical physics gave predicted the spectrum at long wavelengths very accurately, but it also predicted that there was nothing to stop the intensity increasing without limit towards short wavelengths. This embarrassment became known as the ultra-violet catastrophe.

Resolution of these problems of classical physics began to evolve in 1900 with an idea by Max Planck. He showed that the ultra-violet catastrophe could be resolved by assuming that electromagnetic energy came in packets whose size was proportional to the frequency. A wave of a given frequency, f, could not exist with any energy, E — its energy had to be in integer multiples of an energy $E = hf$, where h was a constant, now known as Planck's constant. In 1905, the same miraculous year that he published the Special Theory of Relativity, Einstein analysed the photoelectric effect, the emission of electrons from an illuminated metal, and showed that it could be understood in terms of Planck's quanta of energy, now known as photons. To get an electron out of a solid needed a certain amount of energy. No amount of light with too low a frequency would do, whatever its intensity (barring two- or more- photon processes, which were unknown at the time!). The frequency of the light had to be above a critical amount so that an individual photon, absorbed by an electron, would allow it to escape. This convinced everyone that Planck's quantum was not just a mathematical trick, but that it was a discovery of how nature worked and how the commonsense ideas of classical physics and its deterministic dogma could be plainly wrong. Quantum theory had been born.

If electromagnetic waves had particle-like properties, then particles like electrons may have wave-like properties. Experiments exhibiting electron diffraction showed that this was indeed the case. Hence was born the infamous, non-intuitive wave-particle duality of matter. But, unlike waves of light, matter waves are waves of probability in that their intensity at any point is a measure of the particle being discovered there. In the case of light the intensity of the wave measures the number of photons.

The introduction of probability, as elsewhere in physics, always reflects ignorance. In classical physics we may not know the dynamic state of each molecule in a gas, but we believe its motion is determined by Newton's Laws of Motion. In the case of the electron, things are more serious: the ignorance is fundamental. Moreover, dynamic quantities become related by Heisenberg's Uncertainty Principle which states the uncertainty in establishing the momentum of a quantum particle, Δp, and the uncertainty in establishing its position, Δx, are related, such that Δp multiplied by Δx can never be less than $\hbar$; mathematically $\Delta p \Delta x > \hbar$. (The symbol $\hbar$ (aitch cross) appears very often in quantum theory. It is equal to h, Planck's constant, divided by 2π.) That this is simply what one would expect in the case of a classical wave-packet, a superposition of waves of different wavelength, can be seen from de Broglie's relation relating momentum with wavelength, i.e., $p = h/\lambda$ Rather than work with the reciprocal of the wavelength it is often neater to work with the wave-vector, $k = 2\pi/\lambda$ The Uncertainty Principle can then be expressed simply as $\Delta k \Delta x > 1$, which is the classical expression for a wave-packet. What this describes is straightforward, given the wave-like nature of the electron. If the momentum of the electron is precisely known, this means that it has a precise wavelength. But if it has a precise wavelength, the wave is infinitely long and where the electron is situated in the wave is completely uncertain. Thus, precise knowledge of momentum precludes any knowledge of position, and vice versa; if the position of the electron is known, its momentum is utterly uncertain. Momentum and position, quite independent quantities in classical physics, now become related in complementary fashion. Other closely related dynamic quantities that are similarly complementary in nature are energy and time: $\Delta E \Delta t > \hbar$. If an observation of an electron is made at a precise time, its energy is completely undefined, and vice versa; the longer the electron spends in a given dynamic state, the more accurately can its energy be determined. More disturbing, a photon of energy ΔE can appear out of nothing as long as it disappears quickly enough, so the Uncertainty Principle is not violated. Such a phenomenon is known as a virtual particle. This is a first indication that the vacuum may not be as neutral and placid as classical physics would have it. The void is not a void after all. It is full of virtual particles.

What this calls for is a new concept that is applicable throughout nature. This was provided by Niels Bohr as The Principle of Complementarity:

> At the quantum level, the most general physical properties of any system must be expressed in terms of complementary pairs of variables, each of which can be better defined only at the expense of a corresponding loss in the degree of definition of the other.

The usual examples are the quantities of classical mechanics — momentum/position, energy/time. Another example of a pair of complementary concepts is continuity and discontinuity in the description of jumps of the electron between energy states in an atom. I propose to extend the Principle to a more homely quantum system: Mind and Body. A focus on the neurology of the brain will drive out any attribute of mind; equally, a focus on feelings and emotions will drive out the concepts of matter; but both are essential for describing the system. This is just the quantum theoretic description of Spinoza's Double Aspect theory. We will follow this up later, but already it is clear that the discovery of the quantum seriously modifies what we think of as matter.

Another extraordinary discovery was the phenomenon of non-localisation. Quantum systems that interact become essentially a single system and remain so no matter how far apart they are in space, unless they are disturbed by one or more other quantum systems. Since, according to quantum theory, everything is a quantum system, this implies that the whole world is just one big coherent system. Experimentally, it is a fact that quantum systems that become entangled into what is a single system remain so even if separated by several metres, (even kilometres!) but once there are interactions with the external world the entanglement disappears, in fact if not in theory. There is an analogy with mental processes that has been pointed out by others. Thoughts do not occupy positions in space, whereas classical material events do,

A mathematical equation that described the evolution of matter waves was discovered by Erwin Schrödinger. Schrödinger's equation became the bed-rock of early quantum theory, successfully accounting for the principal features of the spectrum of radiation emitted by that simplest of all atoms, hydrogen, and much more. It did not include the effects predicted by Einstein's theory of special relativity, but in many cases the speeds of electrons were modest, so that omission was not a problem.

Combining Einstein's special theory of relativity into quantum theory produced another strange element of nature, the *quantum field*. Paul Dirac, attempting to describe the relativistic electron using the famous equation of Schrödinger found that the solution described, not a single electron, but a field of electrons and their anti-particles, the positrons. As well as matter, there was anti-matter, the one the destroyer of the other, should they meet. These particles each carried a spin, a quantum of angular momentum that was ½ in units of Planck's constant. This meant that the electron and its anti-particle were tiny magnets. It was discovered that only particles of opposite spin were allowed to occupy the same energy state, their distribution over states obeyed a statistics known as Fermi-Dirac statistics, and all particles possessing half-integral spin and obeying Fermi-Dirac statistics are known as fermions. This included the neutral particle — the neutron — that inhabited the nuclei of heavier-than-hydrogen atoms. The electromagnetic field was already relativistic and could be described in terms of a quantum field of photons. Light can be circularly polarised, and this shows that each photon carries a spin of unity. Since there are no restrictions on the number of photons that can occupy a single state, they obey a different statistics, that of Bose-Einstein. All particles with integer spin are known as bosons. The world is therefore a duality of fermions and bosons. And permeating everything are the quantum fields, in ceaseless turmoil seething, with the continual appearance and disappearance of virtual particles. Aristotle and Descartes were right; there is no void, but, instead, an eternal bubbling of virtual matter.

Protons and neutrons, the constituents of atomic nuclei, each have the same spin as the electron and are therefore fermions, but they are much bigger and heavier particles. Unlike electrons which are, as near as anything, point particles, protons and neutrons have a radius of about 10^{-13}cm and have a structure. The study by high-energy-particle physicists of the galaxy of particles that are produced in special machines, together with the study of radioactivity, which is the spontaneous decay of atoms into other atoms, has led to the so-called *Standard Model* of the elementary constituents of matter.

> 1. *Quarks*. These are the point-like particles that protons, neutrons and other heavy particles are made of. They have a spin like the electron and they come in six 'flavours': up, down, charm, strangeness, truth and beauty. (In this case truth, *pace* Keats, has to be distinguished from beauty, and that is all you need to know.) They also come in three colours.
>
> 2. *Leptons*. There are six types of leptons: electron, muon, tauon plus three types of neutrino, these latter massless (or nearly so).
>
> 3. *Force carriers*. The graviton, guessed-at-quantum of the gravitational field; the photon, established quantum of the electromagnetic field; the $W^{\pm}$ and Z^0 particles of the electroweak interaction (radioactivity); and eight gluons of the strong interaction (intra-nuclear forces).
>
> 4. *Origin of mass*. A field that accounts for why the particles have mass, is associated with the Higgs particle, that is predicted to exist, though yet to be observed.

Then there are the anti-particles.

The quest for enormous energies made it irresistible to combine particle physics with cosmology in an investigation of the earliest moments of the Big Bang when huge energies were surely the norm. The need is for a theory that explains why there are so many particles, why there are four forces — the strong nuclear force, the electromagnetic force, the weak radioactive force, the gravitational force — why is gravity so relatively weak, why are so many basic dimensionless numbers the same, why there are fermions and bosons and how gravity and quantum theory can be made compatible. Grand Unified Theory purports to be that theory but it is bedevilled by infinities and other technical problems, with a hubris all too familiar, The Theory of Everything.

The search for such a theory starts in a promising way by focusing critically on two basic concepts that are common in both quantum theory and in general relativity, the point particle and the continuum. Regarding particles as points is in many ways convenient but it leads to the problem of infinite self-energies that have to be bypassed in quantum-field theory by some sort of mathematical leger-de-main. Seeing particles as extended objects held out the hope that infinities might be avoided. To go from zero-dimensional objects the simplest step is to go to one-dimensional objects, strings. But how long is a piece of string? The natural unit of length suggests that space may be grainy at distances of order of the Planck length, so the length of a string in superstring theory is about 10^{-35}m which is so small that it is not surprising that the idea of a point particle persisted for so long. The string is envisaged to be under enormous tension so it vibrates at high frequencies with vibrating, rotating and twisting mode patterns. Its ends must travel at the speed of light and this means that the string itself has to be massless. In quantum theory vibrations are always quantized and the quantum of vibrational energy is given by $\hbar\omega$, where ω is the angular frequency, which is then related to a mass via $E = mc^2$. In this way a massless string can represent a massive particle and its harmonics. Particle collisions are described in terms of strings breaking and then joining, sometimes forming loops, quite a different image from the usual particle picture. But for strings accurately to describe the particles we know, to be at the same time relativistic, quantized and capable of including gravity, something had to give. In this case it was the old-fashioned four-dimensional world—three space and one time. It was necessary for the vibrations to be in no less than ten space dimensions. Even then there were many anomalies. In an impressive step forward, it was found that these anomalies disappeared for a single choice of symmetry which involves thirty-two dimensions. The theory could then describe fermions and bosons on an equal basis and to contain both quantum fields and gravity, the latter associated with closed loops. At low energies, it is postulated, six of the ten space dimensions become 'compacted' and unobservable, leaving the four of the observable world.

An even more profligate positing of dimensions is to be found in the theory of heterotic strings. These are chosen to be all closed loops that allow waves carrying quantum numbers to travel both clockwise and anticlockwise. In one direction the waves refer to fermions, in the other to bosons, so paired fermions and bosons are neatly unified while being quite separate. The trouble is — and this is where the term heterotic comes in — the fermion waves need the usual ten dimensions but the boson waves need twenty-six dimensions, so now a string exists in *two* different dimensional spaces. It is not clear how in the transition to our world of low energies the dimensions become compacted so only four remain, nor is it clear what these extra dimensions are. The view seems to be that the real world is really a world of many dimensions, but we only see four. The extra dimensions are unobservable. This is a fine example of mathematical myth which sets forth a picture of the universe far beyond the power of empiricism. And, like all theology, it is a Theory of Everything. Truly magnificent! Mathematics par excellence! It may even have some bearing on the real world. Unfortunately, string theory has yet to acquire empirical justification, though it may do so to some extent once the Large Hadron Collider at CERN, the European Laboratory of Particle Physics, in Geneva, is got working again.

But perhaps even more magnificent was what emerged as an aside from CERN, not a new particle, but no less than the World Wide Web, a truly significant contribution to humanity. To appreciate its origins, we have to go back to the Cold War and the perceived military needs of that time.

A huge pressure was the requirement for rapid transfer of information from one computer to another. Schemes were developed in the USA and at NPL (the National Physics Laboratory) in the UK. By the end of 1969 the first network was set up involving four computers—at UCLA, the Stanford Research Instituite, UC Santa Barbara and the University of Utah. So was born the Internet! Other institutions soon joined and, by 1972, that modern phenomenon—electronic mail—was born.

The power of the Internet, particularly in its manifestation as email and rapid access to information, became publicly apparent—there was more for society in this than the exchange of military data. In 1989, a software consultant at CERN, Tim Berners-Lee, conceived the idea of the World Wide Web that was to enable contacts between computers globally. CERN, seeing advantages in facilitating the exchange of technical information, internally and externally, supported the idea. Berners-Lee went on to develop the necessary servers—the gnomic 'http' or 'html' that precede www on our computers—that were eventually incorporated on the Internet in 1991. Today, Berners-Lee, now (worthily) Sir Timothy, directs the development of the Web from his new post at MIT. Curmudgeons apart, most have welcomed the advent of email and the other features of the Web, which have been revolutionary in themselves. In this, particle physics has, inadvertently been substantially beneficial to mankind.

The study of the physics of the atomic nucleus has given humanity, for better or worse, nuclear energy, in its most extreme form, the energy of the stars. On the other hand, the plethora of elementary particles produced in the vast machines of particle physics has given us the Standard Model, but few technological benefits. Instead of revealing the arche of the Pre-Socratics, science has uncovered a vast plurality. Even the simplest form of matter cannot be reduced to less than three fundamental things: the up quark, the down quark and the electron (and to do this we have to ignore the anti-particles and the force-carrying bosons).

Modern humanity's need is not for more particles, but for faster communication. The shorter the wavelength, the higher the frequency, the more information that can be crammed into a small space, and the faster a host of messages can be transmitted. Hence nanotechnology, in which the size of devices get measured in terms of nanometres (10^{-9}m), where a fifth of a nanometre is roughly the size of an atom. Transistors have to get smaller and smaller for computers to work faster and faster, and television to work clearer and clearer. In the 1960s at the beginning of the crystal revolution, Gerald Moore, manufacturing silicon chips in Silicon Valley, discovered a law that stated that the number of transistors fabricated on a silicon chip doubled every eighteen months. At the beginning that number was a thousand: by 2000 it was nearly 100 million, and there is, as yet, no sign of Moore's Law breaking down. This rate predicts that in one hundred years or so the number will exceed the total number of atoms in the universe! Quantum effects are bound to impose a limit. The users of cell phones, ipods, and computers generally, will just have to put up with it. However, it is a burden that most people will shoulder with equanimity.

The science of spectroscopy, beginning around 1800, soon discovered that each element had a characteristic pattern of wavelengths that were absorbed or emitted. Quantum theory explained this in terms of the energy-level structure of electrons corresponding to their allowed wave patterns. Light comes in photons, packets of energy determined by the wavelength. An electron can absorb the energy of a photon only if, by doing so, it can make a transition to an empty, higher energy level. Measuring the wavelengths that get absorbed, and using Planck's constant to convert wavelength to photon energy, reveals the energy-level structure of the atom. When the atoms are hot, the light that is emitted reveals the same structure. Sodium vapour emits yellow, a familiar feature of street lighting, neon emits red, and so on. The bar-code-like pattern of spectral lines is characteristic of the element and identifies its presence.

Adding the telescope and the photographic plate to spectroscopy initiated another new science — astrophysics, the study of the stars. The first discovery was that the sun contained elements that were the same elements familiar on earth; the sun was not as mysterious as it might have been. But there was one pattern of spectroscopic lines that was new, and the new element was called helium. It was subsequently discovered on earth. The material kinship of earth and sun was found to encompass the stars themselves. The spectra of hydrogen and helium were commonly found in starlight, as well as the spectra of many earth elements. This inverted the old Neoplatonic saying to 'As below, so above', but it was an encouragement to astronomers to apply earth-bound knowledge to the stars. The conversion of hydrogen to helium via nuclear fusion, a process achieved on earth, was seen as an important source of stellar power. Stars became objects that could be studied scientifically, how they were born, how they evolved and how they could die. So much for the static Aristotelian Empyrium!

As telescopes grew more and more powerful, the universe grew in size and complexity, and man's place in it seemed to shrink in proportion. Far from the Earth being the centre of the world, it became a planet, one among many, orbiting an unremarkable star situated in the suburbs of the enormous family of stars we call the Milky Way Galaxy, itself a mere member of the local galactic cluster, only one component of a supercluster. As Giordano Bruno had asserted at the end of the sixteenth century, the cosmos contained numberless worlds.

But even the belief in the eternal nature of the heavens, already shaken by Tycho Brahe's new star of 1572, was further undermined by the astrophysicist's account of stellar evolution. Change was everywhere. Spectroscopy discovered that the spectral lines of some stars were shifted slightly towards the blue and some were shifted slightly towards the red. The explanation lay in the Doppler Effect — the frequency of waves, whether of sound waves from ambulances and police cars or of light, increases when their source approaches, and decreases when their source recedes. So the stars were moving relative to us, some coming towards us, some going away, and their speeds could be deduced from the spectral shifts. Similar results were found in the case of nearby galaxies, but for more distant galaxies the shift moved systematically to the red.

The measurement of distance was a problem that was solved by the study of the characteristic behaviour of certain variable stars. Detecting one of these stars in a distant galaxy and assuming its intrinsic brightness was unchanged, astronomers could then estimate the distance. In this, and in other ways, galactic distances could be determined and related to the red shift. Correlating this data in 1929, Edwin Hubble made a remarkable discovery that was to change man's perception of his universe forever. He found that the red shift was directly proportional to distance, and this became known as Hubble's Law. It meant that the universe was expanding, with galaxies receding from us more rapidly the further they were away. It was as if the universe in the distant past had exploded, and galaxies having the greater velocity naturally becoming more distant. Thus was born the Big Bang theory of the universe and the beginning of the new science of cosmology based on Einstein's theory of General Relativity. We return to this topic in a later chapter.

Chapter Ten

Towards a Mathematical Theology

We see in modern physics a shift from the hubris of dogmatic materialism to the hubris of mathematical theology. In order to trace the path along which it has come about, and to make comprehensible why it has come about, it will be necessary to recap some of the details of the progress of physics.

Physics without mathematics is unimaginable; though Francis Bacon, in his seventeenth century clarion call to scientific arms, managed scarcely to mention it. But Kepler's discovery that planetary orbits were ellipses, Galileo's quantitative description of motion and Newton's clothing of the dangerously occult notion of action at a distance with the cloak of field theory, all of these set the mathematical tone of classical physics. Homely things like velocity and acceleration, pushes and pulls, energy and momentum — the stuff of mechanics — part of everyday experience, were incorporated into a mathematical structure of enormous elegance — a jewel of classical physics — by mathematicians like Euler, Lagrange and Hamilton. Classical mechanics generalized the insights of Galileo and Newton to the case of many particles, but it embodied nothing more than Newton's Laws of Motion, which were firmly founded on empirical fact.

The studies of Gilbert, Coulomb, Ampère and Faraday revealed in quantitative detail yet another magical action-at-a-distance, this time between magnets and electric charges. The rational account of this, in the form of the electromagnetic field, was provided by Maxwell. Mathematically more complex than the gravitational field, it nevertheless incorporated Newton's ideas and inhabited every point in space with a potential (for motion). Having to describe both electric and magnetic fields, the potential is a vector potential (magnitude and direction) unlike the gravitational potential, which is a scalar potential (magnitude only). The concept of a vector field, added to the concept of a scalar field, allowed classical physics to describe the motion of matter in all circumstances then known. An additional triumph of electromagnetic theory was its description of that non-material phenomenon, light. Not only could it account for the wave-like nature of light, revealed vividly in Young's famous double slit experiment, it absolutely defined its velocity in terms of the properties of space itself.

In Maxwell's account of electromagnetism another step was taken by physics towards the purely mathematical description of nature. There was first, Newton's action-at-a-distance of gravity, that appalled Descartes among many others, as being nothing more than the sort of occult magical influences forming the core of Rosicrucianism and the Hermetic philosophy in general. Descartes preferred a physical picture, one in which the planets were pushed along in their orbits around the sun by vortices. Magic it may have been, but the mathematics of the scalar field could account quantitatively for the motion of matter in the vicinity of other matter without bringing in vortices. On the other hand, in mechanics, there were no occult influences; just the straightforward pushes and pulls of one bit of matter touching another, which accounted for the laws of billiard balls, water waves and even sound. But electromagnetic waves were something else again. What was vibrating? The answer was the vector potential. Vibrating in what? The vacuum. In

Descartian terms, there is no physical picture — of one thing pushing against another thing. Our understanding is purely mathematical.

For those who prefer, and possibly need, a physical picture of physical phenomena — and that includes many physicists — worse was to come. Space (as distinct from geometry) has always been a problem. Aristotle thought that space was defined by the things that were in it. It seemed absurd to him and others that came after him that space could exist in its own right. Newton disagreed and Kant was convinced that space and time were a priori concepts of the human mind. But if space has absolute attributes then the speed of light, defined by the properties of the vacuum, must be a fundamental constant, the same for all observers. Following the logic of this led Einstein to his Special Theory of Relativity which abolished the Newtonian idea of Absolute Space and Absolute Time and generalized the laws of motion so that all observers, whatever their relative uniform motion, would agree about the laws of physics. Observers travelling with uniform relative velocity with respect to others are known as inertial observers. Any one inertial observer still sees space and time as distinct things, but his interpretation of phenomena going on in another inertial system requires him to see space and time in that system carved up in a quite different way. In the world of special relativity space and time become fused together into a four-dimensional space-time continuum in which relations between inertial observers are mediated by mathematical formulae known as the Lorentz transformations. Moving objects appear to be rotated and moving clocks appear to run slow, but these effects are very tiny unless the velocity is near the velocity of light. Nevertheless, our common sense ideas of space and time have to be modified and our navigation through the space-time continuum has to be done mathematically.

The Principle of General Covariance — the idea that all observers, inertial and non-inertial (i.e., accelerated), should see the same physics — led Einstein via another principle, the

Principle of Mass Equivalence, to the General Theory of Relativity. The mass of a body that appeared in Newton's laws of motion is the inertial mass; it quantifies the reluctance of a body to change its motion. The mass of a body that appears in Newton's theory of gravity is the gravitational mass; it quantifies the gravitational charge that is the source of gravity. There is no obvious reason why these two masses should be the same, but experiments with increasing accuracy over the years have never found any difference. The Principle of Mass Equivalence states that they are identical. If so, then in a local region of space the effect of gravity must be indistinguishable from that of acceleration. A velocity can then be attached to each point in the gravitational field and the laws of special relativity will tell us what space and time looks like at that point as viewed from far away from the gravitational source. In short, space-time becomes distorted in a gravitational field and the geometry of space-time is no longer Euclidean. Adopting the results of Riemann, Lobatchevsky and others who worked out the pure mathematics of non-Euclidean geometry in the nineteenth century, and adding the equivalence of mass and energy that was predicted by special relativity, Einstein could write down his famous field equation that related the geometry of space-time to the energy density of matter and radiation. The concept of a gravitational field became the concept of space-time curvature.

Certain solutions of Einstein's field equation, obtained for simple cases (eg., for a perfectly spherical, motionless gravitating body), make predictions concerning clocks running slow, the bending of light, and the precession of planetary orbits, all of which have been verified by measurement. The connection between the Principle of Mass Equivalence and gravity, that is at the heart of general relativity, is well established. The existence of black holes, collapsed stars so dense that light cannot escape, can be revealed by their gravitational effect and there are several candidates that are observed. Applying general relativity to the universe as a whole brings us to cosmology and with it the germs of a kind of mathematical theology. But before going into that, we can remark that the Theories of Relativity, Electromagnetism and Mechanics complete classical physics, which may be conveniently defined by Einstein's memorable comment about quantum theory, as the part of nature where 'God does not play dice'.

The discovery of quantum phenomena by Planck, and their theoretical description by Bohr, Schrödinger, Heisenberg and Dirac, showed that, in fact, God was not averse to playing dice. Quantum theory placed mathematics at the centre of things with a vengeance. Particles like electrons became waves, but like no other waves known to physics, being defined as complex quantities in the mathematical sense that they contain the square root of minus one, a wholly imaginary number. Contact with reality is in the modulus square of the amplitude, which is interpreted as the probability density of detecting an electron in an experiment. Electrons are certainly waves since they are diffracted like X-rays by the regular atomic lattice of a crystal, but equally they are certainly particles since they can be detected at a particular point on a photographic plate or on a television screen. Light is equally certainly waves, but the photoelectric effect shows light to be particles, photons. Classical solid-state physics describes the motion of atoms in a crystal lattice in terms of waves, but they too are found to behave like particles — phonons. Physical pictures of quantum events become difficult without a kind of double-think involving both waves and particles and the abandonment of the strictly classical relation between cause and effect. Yet all is not chaos: Schrödinger's equation describes the deterministic evolution of the electron wave. What is more, it countenances the superposition of waves so that two electrons become one system no matter how far apart they are. Alain Aspect and his team have verified that this non-locality is true of photons over at least several metres (which others have extended to kilometres!). But there is nothing to explain how the result of a measurement comes about. A measurement that shows that an electron is here rather than there means that a certainty has emerged from a probability, an effect often referred to as the collapse of the wavefunction. But there is nothing in Schrödinger's equation that accounts for a collapse. Any interaction with a measuring system results in a superposition of wavefunctions, but not a collapse.

In most applications of quantum theory, notably in solid-state physics, this paradox is ignored. Indeed, in much of the everyday theory of the properties of crystals, electrons, photons and phonons are envisaged, more or less, as classical particles. Once quantum theory had explained the structure of the atomic bonds that held a solid together, and told how the energy of the electrons in those bonds was ordered into energy bands, the distinction between conductors and insulators began to become clear. Further input revealed the important distinction between classical statistics and the statistics necessary to describe electrons, photons and phonons, and their interactions. The stage was then set for classical ideas, suitably modified, to come to the fore and trigger the engineering revolution that became known as Information Technology. But recent advances in the growth of special structures—quantum wells, quantum wires and quantum dots—remind us that, in spite of apparent classical behaviour, the world is really a quantum world.

Along with the phenomenon of non-locality and the blurring of particles and waves, quantum theory has enlivened the vacuum to a remarkable degree and, furthermore, it has engendered a new type of field—the quantum field. Theory treats the electromagnetic field in a vacuum through an analogy with the quantum theory of simple harmonic oscillators whose energy comes in quanta. The more energy in a wave of a given frequency the more quanta (photons) are present and the classical energy of the wave can be understood in terms of so many photons. Nothing much is new until one thinks of reducing the energy of the wave to zero. It cannot be done.

There is always a so-called zero-point oscillation of the simple harmonic oscillator which contributes exactly one half of a photon's energy. It is as if a photon exists for half the time and vanishes for half the time. But this is what the Uncertainty Principle predicts: an uncertainty in energy multiplied by an uncertainty in time can be of order of Planck's constant. The vacuum has now to be seen as bubbling continuously with photons leading a transient existence, even when there is no real electromagnetic field there. Such photons are called virtual particles to distinguish them from the real particles that exist when a real field is present. Heinrich Casimir predicted that the existence of these virtual photons would mean that two metallic plates would attract one another, and this has been amply confirmed. The bubbling vacuum is real.

One of the delightful properties of the quantum vacuum is that it can account for that notorious action-at-a-distance problem. Waves of light are always transversely polarized, that is, the electric and magnetic fields of the wave are directed at right angles to the direction of propagation of the light. Longitudinally polarized photons do not exist except by virtue of the Uncertainty Principle, which allows them the usual brief existence of a virtual particle. Action-at-a-distance can be understood as the exchange of virtual, longitudinally-polarized photons. Photons being massless allow the range of the interaction to be unrestricted. Their finite speed means that the interaction cannot be instantaneous but, rather, retarded. A similar explanation for gravity requires the existence of gravitational waves and the exchange of gravitons. Although a completely satisfactory quantum theory of gravity has yet to be formulated, the existence of gravitational waves is taken seriously, and attempts to detect their existence are continually being made. The observed decay of the orbits of a binary pulsar is evidence that gravitational energy is being radiated away via gravitational waves.

The electromagnetic field already existed before the discovery of quantum effects, so its conversion to a quantum field was straightforward. A corresponding quantum field for particles did not exist until Dirac's analysis of the special-relativistic version of Schrödinger's equation. Changing Schrödinger's equation to agree with special relativity meant introducing rest-mass energy. The result is the Klein-Gordon equation. It describes a scalar field for particles with mass but with no internal motion, the latter known as spin. Particles with spin that is a whole number times the quantum of angular momentum are known as bosons and obey Bose-Einstein statistics. Photons, being products of a vector field, are spin 1 bosons. The particles described by the Klein-Gordon equation are spin 0 bosons, which do not appear to exist, sadly enough. Particles like the electron have half integer spin and are called fermions, obeying Fermi-Dirac statistics. Dirac's achievment was to find a factorized version of the Klein-Gordon equation that could describe electrons in terms of what is known as a spinor field. In addition, the negative energy solutions of Dirac's equation could be interpreted as anti-particles — in the case of the electron, the positron. The discovery of anti-particles raised the question of why they do not seem to exist in the universe as abundantly as particles, given their similarities. There is as yet no answer to this. But, quite apart from that, it soon became clear that, unlike the Schrödinger equation that described a single particle, the Dirac equation was really an equation for many particles, the quanta, in a word, for a field. As in the case of the electromagnetic field, the field described virtual as well as real particles, the virtual particles coming as virtual electron-positron pairs (conserving charge). Virtual particles jostle with virtual photons in the increasingly active vacuum.

Rather as the Newtonian mechanics of a few particles evolved into the Lagrangian and Hamiltonian mechanics of many particles, the Schrödinger equation has evolved into the quantum field. The basic particles of the Standard Model, quarks (fermions) and gluons (bosons), must now be seen as creatures of their individual quantum fields. In his paper in which he predicted the existence of magnetic monopoles, Dirac advocated tackling the difficult problems that nature throws up by employing

> all the resources of pure mathematics in attempts to perfect and generalise the mathematical formalism that forms the existing basis of theoretical physics, and *after* each success in this direction, to try to interpret the new mathematical features in terms of physical entities. [The italics are Dirac's.]

Magnetic monopoles have never been seen. However, the manipulation of quantum field theories by Weinberg and Salam have yielded a fusion between electromagnetism and the so-called weak interaction that mediates radioactive decay. They predicted the properties of the related bosons (W and Z), and these have been observed. The creation of a new quantum field — the electroweak field — is a fine example of Dirac's methodology.

A vital component has been the concept of symmetry. This idea is, of course, a familiar one in solid-state physics where crystal classes are defined in terms of the spatial operations that can be performed (rotations, reflections, etc.) without changing the crystal lattice. In group theory all the possible symmetry operations are codified and their importance in physics is their connection with what physical processes are allowed and which are not. Of even more importance is that a study of symmetry can define quantities that remain invariant. If nothing changes in a system that is translated in space, then it turns out that momentum must be conserved. If there is in the equations time-reversal symmetry, energy is conserved. Symmetry is therefore a powerful tool in ordinary physics for identifying invariants, but when it comes to inventing new quantum fields it has to be generalised beyond spatial operations. Thus, in electromagnetism, if there are no changes when the zero of electrostatic potential is shifted, then charge is conserved.

In electromagnetism the conservation of electric charge is basic and can be thought of as a result of the symmetry associated with potential. But by conservation of charge we do not mean that if 10 coulombs disappear in Colchester 10 coulombs must appear in Newcastle or on the moon. That sort of symmetry implies that we could shift every point in space the same amount, which amounts to changing the measure or gauge, and everything would remain the same. It would be a global gauge symmetry whose instantaneous operation would defy the tenets of special relativity. The symmetry that determines charge conservation must therefore be local and it must obey the laws of special relativity. The consequences of demanding local gauge symmetry are very far-reaching. Whatever the symmetry refers to it must be maintained going from a given point in space to an adjacent point infinitesimally distant. This implies the existence of forces that ensures that this symmetry is maintained everywhere without violating special relativity. In other words, local gauge symmetry always entails the existence of a field with quanta that are massless. The exchange of virtual particles that constitute force means that, consistent with the Uncertainty Principle, the particles cannot have mass if the field is to be of infinite range. In formal language, such a field must guarantee invariance everywhere under local gauge transformations. The electromagnetic field is exactly that sort of field and the abstract symmetry involved is known as $U(1)$.

Gauge theory takes the electromagnetic field as the fundamental exemplar, which entails that all quantum fields must have this property of guaranteeing invariance under local gauge transformations while obeying special relativity. Such fields are known as gauge fields. In the simplest, the electromagnetic field, the force between two charged particles is carried by the photon. The interacting particles do not change—electrons remain electrons, photons remain photons—and the abstract symmetry is the unitary group in one dimension, $U(1)$. The weak force, the force responsible for radioactive decay, is more complicated because particles *do* change. A neutron colliding with a neutrino can change into a proton while the neutrino changes to an electron. Since charge is transferred the quantum carrying the weak force must carry a negative charge. This is the W^- particle. Reactions involving antiparticles will be mediated by the positively charged W^+ particle and reactions where no charge is transferred, for example between an electron and a neutrino, will involve the neutral Z particle. The symmetry associated with the weak force is the special unitary group (special because its matrices have unit determinant) in *two* dimensions, SU(2) and its quanta are known as intermediate vector bosons. The electroweak force has symmetry combination SU(2)xU(1).

This is all fine, but, unfortunately, the weak-field bosons are found to have mass, which renders the interaction short range. This violates the condition for the weak force to be describable in terms of a gauge field, which must have infinite range with, therefore, massless quanta. To save the fusion of the weak and electromagnetic interactions there is evoked the concept of broken symmetry. In the purely symmetric state the weak-force particles, like the photon, are massless and therefore describable by a gauge field.

It goes like this. Esoteric manipulation of the quantum field concept leads to the ideas of a false vacuum and spontaneous symmetry breaking. In the case of ordinary, standard fields the field amplitude vanishes when the energy is zero (ignoring zero-point effects) and that defines the vacuum state. Regarding the rest-mass term that appears in formulations like the Klein-Gordon equation as just a parameter and nothing directly to do with the mass of a particle we can take it to be negative (why not?) as if the mass was purely imaginary. The result is a so-called Higgs field and the creation of a false vacuum in which the field amplitude vanishes at a finite energy above zero. This point is nicely symmetrical in Higgs-field space, but being at a finite energy above zero the situation is metastable. Just as a pencil balanced carefully on its point must ultimately fall flat, the Higgs field must ultimately convert to the less symmetrical state in which the field amplitude is finite but the energy is zero. This is what is meant by symmetry breaking. In this process, known as the Higgs phenomenon, the particles in the symmetrical state disappear and in the true vacuum new particles appear. In electroweak theory a scalar field, the Higgs field, is added to the weak and electromagnetic components and when symmetry is broken the weak-field bosons acquire mass. There should be several Higgs fields and several associated bosons, but search for the least massive of these has so far failed, a failure usually attributed to its mass being too large for existing machines to detect. There is clearly dangerous metaphysical ground here, but it would be premature and even joyless to equate this ingenious theory with a kind of mathematical mythology.

The Standard Model of particle physics encompasses no less than thirty-seven particles, not counting anti-particles. As regards fermi particles, there is the electron and there are the electron-like particles, the muon and taon, plus their associated neutrinos, known collectively as leptons, six in all; there are six quarks each carrying one of three colour charges, making eighteen. As regards bosons, there is the photon, there are the three 'intermediate bosons' for the weak interaction and there are eight colour-charged gluons, making twelve in all, or thirteen if one counts the Higgs boson. All told there are thirty-seven particles with a range of masses and attributes. The idea of spontaneous symmetry breaking is attractive because it suggests a mechanism akin to the Fall from Grace whereby this untidy plethora of particles has been generated from a few particles or even a single particle that existed once at enormously high energies. Here is a modern mythology of some power.

A Grand Unified Theory is naturally envisaged that would describe both quarks and leptons by a gauge field incorporating the U(1) symmetry of electromagnetism, the SU(2) symmetry of the weak interaction and the quark SU(3) symmetry of the strong interaction, with the electroweak interaction united with the strong interaction at enormous energies. But beyond this there is the stupendous idea that a state of supersymmetry exists in which fermions and bosons are united, in the sense that every fermion would have a boson counterpart and vice versa. An electron boson would be the selectron, a photon fermion would be the photino, and so on, doubling the number of particles (none of which has been seen), but achieving a satisfying symmetry. Doubly satisfying, in fact, because virtual bosons have positive energy and virtual fermions have negative energy, so if the vacuum is to remain gravitationally neutral it is important that these energies cancel exactly. The gauge field of supersymmetry should turn out to be nothing more than the gravitational field itself. If all of this could be established, Grand Unification would indeed have been achieved. The enormous energies required for supersymmetry are just what are predicted by the Big Bang theory.

The quest for enormous energies, where what passes for Grace in physics resides, made it irresistible to combine particle physics with cosmology in an investigation of the earliest moments of the Big Bang, when huge energies were surely the norm. The need is for a theory that explains why there are so many particles, why there are four forces — the strong nuclear force, the electromagnetic force, the weak radioactive force, the gravitational force — why is gravity so weak, why are so many basic dimensionless numbers the same, why there are fermions and bosons and how gravity and quantum theory can be made compatible. Grand Unified Theory purports to be that theory but it is bedevilled by infinities and other technical problems. String theory claims to be free of these difficulties but at the expense of postulating extra space dimensions. We have a theory of gravity in the form of general relativity, but no quantum gravity.

Hawking and Penrose have shown that space-time singularities are endemic in general relativity. In other words there are points where space-time curvature blows up and becomes infinite and such points therefore cease to be properly part of space-time. A point that is outside of space-time is also outside the laws of physics, and anything may happen there. The Big Bang is one such singularity; fortunately it is so far in the past that is unobservable, though its effects are certainly observable. Another such singularity lies at the centre of a black hole, but again, happily, we cannot penetrate beyond the event horizon, so it remains hidden. One that is not hidden is known as a naked singularity. Penrose has speculated that, being literally beyond physics, naked singularities are banned by a kind of cosmic censorship. Thank God?

Basically, it seems that general relativity, by coming up with geometric blow-ups, is defining limits to its own writ. It may be, however, that incorporating quantum theory may solve the problem of singularities, though, given the delicate treading around infinities in quantum-field theory that has to be done, one may be forgiven for feeling pessimistic. The move away from points to extended structures that occur in string theory may indicate a direction that is more promising. Enter p-branes. If strings are one-dimensional then they are just one example of set of structures called p-branes, where p is the number of dimensions. Thus, strings are 1-branes, but what about 2-branes? If one of those ten or eleven or more dimensions that enter string theory is not curled up microscopically in our familiar 4-dimensions as they are thought to be, but somehow extended orthogonally, it would mean that our universe was like a 4-dimensional membrane in a larger dimensional universe. The idea is that only gravity knows about this larger dimensional universe, which ought to show up in its finer-grained properties. Maybe we live in (or on?) a 2-brane, and there may be neighbouring branes that will have an effect of gravity. The good thing about these apparently crazed ravings of theoreticians is that there is the possibility of experiments on gravity testing predictions that may be made on the basis of p-brane theory.

The mathematical ingenuity displayed in particle physics, quantum-field theory, string theory and cosmology is awe-inspiring. What now is sorely needed is some new observation that bears on these matters. We need to discover the Higgs particle. We need to demonstrate the existence of a false vacuum. We need to see a supersymmetric particle. We need to show that more than four dimensions exist. We need to show that gravity has hitherto unobserved fine-grain properties. The monism that brings gravity into the quantum fold will be another awesome mathematical tour-de-force, but it had better include a suggestion for a do-able experiment to test it if it is to give us confidence that it is still physics, not another example of mathematical theology.

But, hey, physics isn't everything. Surely, the creation itself of imaginative new worlds is intriguing enough. Distinguished from science fiction by its mathematical character, this new activity may be seen eventually as a novel adornment to our culture. Its models of universes, unhampered by the necessity of empirical verification, judged purely by their originality and mathematical elegance, presage a new art-form and a new topic in aesthetics. Our current cosmological visions may be still enmeshed in physics, but the bonds appear to be getting weaker by the decade. What mathematical forces may shortly be released!

One might distinguish two competing strands, as old as philosophy. There is monism and there is pluralism, the one like a religion that urges a faith in one God, the other needing a profligation for theory's sake. Commonplace pluralism might accept with equanimity the existence of dozens of elementary particles, but monism will only settle for one God-particle from which all others are derived. Theoretical pluralism requires endless duplication either for conceived consistency, as in the Many Worlds scenario of quantum theory, or in string theory regarding the number of dimensions. It may also need proliferation for statistical reasons, as in the Many Universes idea, given that nothing scientifically germane can be said about the unique object our Universe appears to be. Or the cause of empirically-free theories could be neither of these, but merely the good old-fashioned desire to tell a stirring tale. But without empirical support it is not going to be physics as we know it. However, is there not the possibility of empirical support to come, if it be not now? The readiness is all.

In the meantime, it is vital that belief in the correctness of any of these mathematical theories be firmly provisional. Science does not wish to see the establishment of the Church of String Theory, or the Temple of P-Branes, with their associated mathematical theologies distracting too many young scientists. It does not wish to see the great men and women in the field become bishops and gurus whose pronouncements are questioned at the peril of a career in science. Science has to retain its ethos of anything goes that is rational and empirically refutable. Mathematics is certainly rational. Without defining doable experiments it becomes a theology or, at best, a conscious art-form.

Chapter Eleven

Origins

Here are three religious questions: why are we here? Why is anything here? Where did we come from? These are questions that, sometimes, in idle moments we ask ourselves, and have asked ourselves, off and on, since childhood, though with increasing pessimism as to any possibility of any satisfactory answers. The religious institutions will tell us that it is all God's work and ineffable. Science will tell us that 'Why' questions are beyond its brief, whereas 'How' questions might be answerable. Since institutional religion is incapable of satisfying that part of the soul that is awkwardly, but perennially, curious, it might be more interesting to rephrase the questions less attractively: how are we here? How is anything here? Where did we come from? And turn to science and theology.

In Genesis, the answers to all three questions are clear — the Earth and everything in it was created by God. On one ecclesiastical count, creation took place some 4004 years before the birth of Jesus. All living things so created were so beautiful and intricate that they were living evidence of a creator. So claimed William Paley (1743–1805) and used that claim as an argument for the existence of God — the argument from design. The delicate organs and mechanisms that made up the human body, or indeed any living form, pointed to the existence of a Maker, someone who has designed and constructed the finished product. Surely, his analogy went, if one were to stumble across a watch for the first time and examined its intricate mechanism of springs and balance wheels, one would certainly conjecture that the watch was a product of a watchmaker who, therefore, necessarily existed. How otherwise to account for such craftsmanship? There was a further truth: all the life-forms of the Earth were the one-off creation of God. The implication was that the origin of the species coincided with the Creation and each has remained unchanged since that time. Up to the beginning of the nineteenth century, that was the accepted dogma.

The first discontented rumblings came from the fledgling science of geology. There were two main bothers, one arising from the study of rocks and landscape, the other from the fossil record. The former screamed about the existence of time epochs hugely longer than a few thousand years. The latter turned up abundant evidence of a record of life-forms that once existed, but which are now extinct. Creationists retort that God created the fossils as well as the rocks and the landscape, as they are now, so what is the problem? Unfortunately for creationism, that effervescence of the mind that craves sufficient reason for things that happen, began to drown the foundations of institutional belief and to release a stream of atheism that threatened belief in the very existence of God. The first to argue against the Biblical cosmologies was James Hatton (1726–97) in his book *Theory of the Earth*, published in 1785. 'No powers,' he said, 'are to be employed that are not natural to the globe, no action to be admitted except those of which we know the principle.' The stratification of rocks with their fossils was not something that happened once and for all, but were processes active today. As geological science advanced, Sir Charles Lyell published his book *The Principles of Geology* in 1830, the stream became a river as more and more evidence turned up that showed the earth was enormously older than was once thought, and that if God created all animals, he had allowed some of them to die off. The life of a species was far from being guaranteed. What was God up to?

The idea that living things were not given once and for all but perhaps had evolved over time was in the intellectual air at the beginning of the nineteenth century. We have already noted that the idea of evolution had already occurred to Empodecles ages ago. Sir Charles Darwin (1809–82) provided an immense amount of compelling evidence for evolution in his book *The Origin of Species* published in 1859. Furthermore, he advanced an explanation of the mechanism whereby evolution came about, namely, *natural selection*. Natural selection was basically nature's copy of the eugenic artificial selection that farmers do to improve their cattle. It differed in that eugenics was not the motivation but merely the survival of the fittest. In nature, the fittest, most attuned to the environment they inhabited, survived, the rest tended to die out. How attunement to the environment was achieved was by gradual adaptation through a natural, random process of variation.

But for evolution to work in this way, adaptation had to be inheritable; successful variation had to be passed on to the offspring of the fittest. What exactly was passed on from parent to child. Jean Baptist de Lamarck (1744–1829) believed that it was the experience of the successful parent, the habits and strategies it had developed, that was passed on. Darwin did not rule this idea out, but did not have a plausible alternative to offer. As it happened, an alternative was at hand. The patient studies of heredity in peas and other plants of the monk Gregor Mendel (1822–84) (yet another example of an ecclesiastic contribution to science) revealed that there were special factors that were transmitted in the process of reproduction. Certainly not experiences or habits, but it was not until 1926 that Thomas Hunt Morgan identified these 'special factors' as *genes*. These were structures physically located on tiny rod-like inhabitants of the cell called *chromosomes*. The subsequent discovery in the nucleus of the cell of deoxyribonucleic acid (DNA) by Meister and the elucidation of its double-helix structure by Francis Crick and James Watson defined the gene as a region of molecular sequence in the helix. So was born the science of genetics.

Genes determined the colour of the eyes, the colour of the hair, the shape of the nose, etc. Adaptation could come about by some mechanism of mutation, some error in the replication of structures in cell division, a random fluctuation of molecular structure, a virus or caused by the bombardment of cosmic rays. Most mutations will inevitably be disadvantageous, on the rule that if it isn't wrong, don't fix it. But occasionally, one mutation will give the individual an edge in the battle for survival, and, if inherited, will give his progeny a similar edge. Too big an advantageous mutation may mean that the girl next door may think you are a freak, and you may find it difficult to find a mate. Changes, therefore, have to be small. Darwin certainly thought so, but gradualism of this sort gives rise to some problems, as we shall see.

Darwin's theory of evolution and its mechanism of gradual adaptation and natural selection is rightly regarded as one of the greatest products of science, on a par with the creations of Newton, Maxwell and Einstein. Few doubt the truth of its basic description of the origin of the species. So does this answer our question 'How are we here?' The Darwinist answer is that we have evolved over hundreds of thousands of years from earlier primate forms, and these, in turn, have evolved from branches that include the apes. Similarly, other vertabraic animals have evolved from more primitive forms. At some point in the distant past vertebrates departed from invertebrates, plants from animals, everything from single cells. And where did these unicellular creatures come from? In corners of the hot, cooling earth there would be favourable places for the right sort of molecules to get together in larger units, some of which turned out to be self-replicating. In this vision, life emerged from the proverbial primeval soup, and we are the heirs of those living croutons.

The Darwinian account of our origins is clearly consonant with materialism. We are creatures of an immensely long chain of random mutations, with no purpose and no meaning other than the fact of our existence. At some stage the material mind appears, but this, as we know, is mysterious. There are, naturally, energetic objections to this view from the religions which are well-known. There are first-rate accounts in the books *Science and Creation* (1988) and *One World* (2007) by John Polkinghorne, an Anglican priest and former Professor of Mathematical Physics at Cambridge. A robust attack on the materialism of fundamentalist Darwinism can be found in the book *Dawkins' God* (2005) by Alister McGrath, Professor of Historical Theology at Oxford. There are, however, scientific objections that need looking at.

According to Darwin, evolution should be expressed in the fossil record in terms of a gradual transition from one species to another, but this is rarely found. At the time Darwin put this down to a fossil record that was woefully incomplete, but this cannot be a plausible explanation today. In fact, what is most commonly observed is the abrupt appearance of a particular form, something that is referred to as *saltation* (from the Latin *saltus*, jump); and a saltation is commonly followed by a long period in which there is no significant change whatsoever, a condition referred to as *statis*, often followed by rapid extinction. In his book *Punctuated Equilibrium* (2007), Stephen Jay Gould claims that saltation followed by stasis is the dominant type of evolution rather than Darwinian gradualism. He is also critical of the common perception that evolution exhibits a history of life as a sequence of forms of increasing complexity. One such sequence might be, for instance, bacterium, jellyfish, trilobite, eurypterid, fish, dinosaur, mammoth, man. While some of this sequence fits the common perception, the case of bacteria certainly does not. Bacteria, appearing some 3.5 million years ago, are still around today with practically zero change and exhibiting not a wit of an appetite for complexity.

> We live ... in a persisting 'age of bacteria' — the organisms that were in the beginning, are now, and

probably ever shall be (until the sun runs out of fuel), the dominant creatures on earth by any standard of evolutionary criterion of biochemical diversity, range of habitats, resistance to extinction, and perhaps ... even in biomass.

Bacteria, it seems, have always been happy with their lot. Primates, from apes to hominids, neanderthals to man, seem not to be happy with their lot, having evolved comparatively quickly.

Modern Darwinism, or Neo-Darwinism as it is often called, has evolved certain characteristics that remind one of a religion riddled with sects. Gradualists versus saltationists, microsaltationists versus macrosaltationists, believers in a purely terrestrial source of life versus believers in a extra terrestrial origin, yet all of them believing in the basic idea of evolution, at least for many organisms. The problem for mainstream Neo-Darwinists, who are gradualists, is to explain how different species evolve from a common ancestor. One answer to this significant problem exploits the idea of communities becoming geographically split by the geologically sudden appearance of a mountain range or a large lake or a desert. The discovery of plate tectonics and the phenomenon of continental drift explained many oddities of biogeography, for instance, why marsupials are found mainly in Australia. Oceanic islands, remote from any continent, have their own idiosyncratic evolving flora and fauna, remarkable for the lack of native mammals, birds and insects and seeds being the only things that could manage to alight on the island. Geographical features certainly play their part. Mutations proceed in the divided communities but there is now no contact between them, so quite different end-points can be reached by gradual changes. Saltationists point to the glaring gap in the fossil record between the Pre-Cambrian and Cambrian eras. Cambrian strata, some 550 million years old, show the major invertebrate groups—trilobites, lamellibranchs, ammonites, etc.—all in an advanced state of evolution, but there is no trace of ancestral life forms from which they have evolved. It may be that the ancestors were squashy life-forms that were too gooey to become fossils, and it may be that proto-trilobites and the rest may yet be discovered. Many life-forms can be traced back millions of years but without revealing connections between different classes and orders. Each mammalian order can be traced back for about 60 million years before vanishing. The many mysteries include the event that wiped out the large dinosaurs. Jumps, rapid in geological time that create new orders, could be a consequence of macromutation, a viewpoint strongly held by some.

The cosmologist Fred Hoyle (1915–2001) likens gradualism to a river that, unlike other rivers, flows gently downhill without the hint of rapids or waterfalls, and regards this as highly improbable. In *Mathematics of Evolution* (1999) he enters the fray of evolutionary statistics dominated by the received wisdom of J. B. S. Haldane and R. A. Fisher and, as a formidable iconoclast in this and in his own field, he is not upset when some of his results disagree with conservative thought.

> Of the books, I would like to recommend especially R. A. Fisher's book A General Theory of Natural Selection for its brilliant obscurity. After two or three months of investigation it will be found possible to understand some of Fisher's sentences.

Hoyle offers a transparent mathematical proof that point mutations among a fixed aggregate of genes cannot produce any positive evolution whatsoever if the process of reproduction is asexual, such as the binary fission type, as is the case of bacteria. This is simply because rarer advantageous mutations tend to be swamped by the more frequent deleterious sort. This would be a recipe for ultimate extinction through adverse environmental conditions unless the organism could somehow hibernate for a while. Bisexual reproduction, on the other hand, does allow positive evolution, but only of a gradualist sort. Hoyle was convinced that new species could come about only through what he called genetic storms, the equivalent of the river reaching Niagara. His work in astronomy and the study of inter-stellar dust suggested the possibility of an extra-terrestrial origin of primitive living organisms. Yet another Hoylian heresy.

So, how are we here? Are we here as a result of a mind-boggling process that has converted a fortunate concurrence of the right sort of molecules thousands of millions years ago in the right sort of earthly or starry environment into a cell and ultimately into self-conscious human being? How did the simplest cell, the sort without a nucleus, that bacteria are made of, come about? How did the more complex cell, one with a nucleus and a host of add-ons, the sort that we and the rest of the flora and fauna are made of — how did that come about? The natural feeling of the improbability of all this being true has to be countered by the scientific knowledge that the earth is 4000 million years old, time for untold generations to experience the effect of mutations of one sort and another.

That sense of improbability is generated not only by the claims of Darwinian evolution regarding the origin of species, there is also the awesome wonder of the cell as revealed by cell biology. Each one of us is a colony of mutually supportive cells, living entities that somehow have come together to form a human sapiens, most of them tiny, some of them — the nerve cells — not so tiny, each of them feeding on the flow of oxygenated blood. Every cell (creature, why not?) has a life and a structure of eye-watering complexity, each knowing its place and organizing within itself the manufacture of proteins as and when Her Highness the DNA in its nucleus offers her expertise. They are creatures that have captured other creatures, such as the one-time bacteria in their new guise as the energy-giving mitochondria, without which the spectacular chemistry of protein generation and the electro-chemical production of nerve signals simply could not happen. In the embryo, cells get to know who they are, what their special task is, and where to go to do it. Mere looking at the huge length of the human genome cannot tell us how all this is done. The more one learns of the complexity of the human body, the more awesome seems its origin. Never mind musing on the origin of species, muse on the origin of cells and their colonies. And to balance that, muse on the literally unimaginable period of time that is the four-billion-year life of the earth. Is it then possible, having balanced evolving complexity against time, to come down in passionate favour of either Darwinian evolution or Intelligent Design? Surely, no fervent belief one way or the other is sensible. It is literally beyond belief.

Whatever the mechanism of evolution favoured by the Neo-Darwinists, it is founded on the belief in the causal efficacy of chance events. Bergson, for one, found this incredible as a complete explanation. He pointed to the wonderful complexity of the human eye. How could that have evolved through a sequence of random events? Suitable formations of the skull to accommodate the eyes; the development of a cornea, a lens, an iris; the production of the retina and its communications — a mini-brain in itself; the optic nerve and appropriate regions of the brain. It all had to come together coherently. To see this miracle as a product of gradualism, he thought, is simply not plausible. To see it as a product of one of Hoyle's genetic storms is scarcely more plausible. Moreover, and here Bergson believes he has the evidence that sinks materialistic evolution forever, the eye with its immense complexity is found in a totally different species, namely, the molluscs and, specifically, the octopus. Here the phenomenon of light is the only common environmental feature. How is it possible that the same complex organ can have developed in two hugely dissimilar organisms as man and mollusc? Not by chance, surely. There has to be a common cause.

Bergson finds this common cause in an impetus for life that affects all evolution, a continuing creative becoming. He calls this impetus the *élan original* of life, inscrutable to intelligence and to be apprehended only by intuition. This élan is not only responsible for all forms of variation in evolution, it is at the heart of all living things. As such it is a reality inevitably beyond science.

In order to get across the flavour of the heretical theory of evolution he proposes, in simple terms, let us follow Bergson and consider the story from the point of view of the *élan* as life.

> The resistance of inert matter was the obstacle that had first to be overcome. Life seems to have succeeded in this by dint of humility, by making itself very small and very insinuating, bending to physical and chemical forces, consenting even to go a part of the way with them, like the switch that adopts for a while the direction of the rail it is endeavouring to leave. Of phenomena in the simplest forms of life, it is hard to say whether they are still physical and chemical or whether they are already vital. Life had to enter thus into the habits of inert matter, in order to draw it little by little, magnetized, as it were, to another track. The animate forms that first appeared were therefore of extreme simplicity. They were probably tiny masses of scarcely differentiated protoplasm, outwardly resembling the amoeba observable today, but possessed of the tremendous internal push that was to raise them to the highest forms of life. That in virtue of this push the first organisms sought to grow as much as possible, seems likely. But organized matter has a limit of expansion that is very quickly reached; beyond a certain point it divides instead of growing. Ages of effort and prodigies of subtlety were probably necessary for life to get past this new obstacle. It succeeded in inducing an increasing number of elements, ready to divide, to remain united. By the division of labour it knotted between them an indestructible bond. The complex and quasi-discontinuous organism is thus made to function as would a continuous living mass which has simply grown bigger.

Life creates organisms in all directions, from unicellular forms to the division of vegetable and animal. Plants are immobile and have no need to move in order to feed; they get their nutrition from the soil and the air. Animals, on the other hand, have to move in order to feed, which calls for more consciousness than is the case for plants. A further division distinguishes animals, like insects, where instinct is dominant, from vertebrates, like mammals, in which consciousness appears with increasing strength, supplementing instinct. In man, conscious intelligence dominates all external action; the internal action of breathing, digesting, feeding and maintenance is largely left to the unconscious instinct. Intelligence and the need for men to move and act upon the matter of the world has given us science, with its detailed and quantitative models of the material universe. The raw force behind all of life cannot be described and understood by the intelligence, by science. For that, it is necessary to consider instinct, which embodies the true evidence of the *élan vital*, and for that, in turn, it is necessary to elevate intuition over intelligence.

It might be reasonably expected that the Human Genome Project launched in 1991, in which the strand of DNA was to be analysed to identify genes, would identify elements that determined whether an embryo would develop into a human being, or into a mouse, or a chimpanzee. Unfortunately there is, as yet, no indication of how that comes about. We mammals have very similar genetic make-up; there is only a relatively tiny difference between us and a mouse. What in our DNA codes for human, and not mouse, remains a mystery. What makes an injury to a bone or organ in a particular organism heal itself into a cured bone or organ of that particular organism remains inscrutable. How does the human body know that it is a human body? There is a holistic element in the genome that has not yet been identified. Until it is discovered, or indeed until life is created in the test tube, Bergson's hypothesis of the *élan vital* remains viable, *pace* fundamentalist Neo-Darwinists.

In the form of the *élan vital*, Bergson is proposing nothing less than the existence of a force of nature as real as gravity and electromagnetism. It is a force that is life itself, a force that drives evolution in an endlessly creative way, but without aiming for a final product. In this it is like the artist who works without a planned outcome, and knows what he has created only when he has created something. It is unlike the craftsman who knows from the outset what he is making, and knows when he has succeeded. In Bergson's view of evolution, there is no idea of an aim to be reached, a zenith to be obtained. Instead there is endless creation. It has to be said, that a force that is not quantifiable like gravity and electromagnetism is not a concept that is attractive to science, nor to the Anglo-Saxon mindset, which may account for the fact that Bergson's ideas are taken more seriously in continental Europe. But, if Bergson is right, the science of evolution does not have the monopoly of truth. His idea of continuous creation (which, incidentally, precedes the cosmological equivalent championed by Hoyle) is, of course, heretical. It is like a force that energizes Plato's Plenum Principle directed towards organic life: the creation of all possible life-forms. It is consonant with Spinoza's God as verb and not noun. It is certainly more interesting than Intelligent Design, the bête noir of Darwinism. But it cannot be tested. It must remain as a myth. But no worse a myth than any in modern science.

Chapter Twelve

The Big Bang Story

Let us now turn to the second question: 'How is anything here?' For this we have to turn to cosmology, a comparatively young science that began in 1917. Its antecedents lay in the work of Maxwell and his discovery that the velocity of light was not like familiar velocities that are always velocities relative to something; this velocity was an absolute constant of nature. It meant that no matter how fast he himself was travelling with respect to something else, an observer would always measure the same speed of light. This paradox was solved by Einstein in 1905 by his *Special Theory of Relativity* in which space and time became merged into a single coordinate system of space-time that was carved up into space and time in a way that depended on the observer and his relative speed. Since electromagnetic waves were the means whereby we observed the world and built up our science, it meant that observers travelling with different velocities relative to one another would map the world differently. For example two events that to one observer appear to happen simultaneously would appear to be separated in time to another; lengths in moving systems relative to an observer would appear shortened, clocks would appear to run slow. All of these strange effects are well understood by the Special Theory, and in any case, become important only when studying particles that travel near the speed of light. The effects on us, associated with the fastest way we get around, i.e., jet travel, are insignificant.

Turning to the question of accelerated systems Einstein saw that a special set of these systems would be indistinguishable from the effects of gravity. A person in a closed lift accelerating upwards would feel his feet pressing against the floor in just the same way that he would if the lift were stationary and he was in a gravitational field. This was an example of the Principle of Equivalence, which codifies the observation that the inertial mass (the mass involved in acceleration) and the gravitational mass (the source of gravity) of a body were identical, and Einstein saw that in place of gravity one could substitute an appropriate acceleration, but this meant, via the Special Theory, that space-time became modified by gravity. Thus was created, in 1916, his *General Theory of Relativity* which described how space-time was affected by matter and energy. In the following year, 1917, Einstein applied his theory to matter on the grand scale in a simple model of the universe, and cosmology was born.

Many models by others soon followed. At the time it was generally thought that the universe was an unchanging thing, but in 1929 Edwin Hubble (1889–1953) astounded the world with his discovery, after years of observing distant galaxies, that the universe was expanding. The same elements composed the stars as on earth. Measurements of the spectra from nearby stars showed shifts in their wavelengths consonant with their motion via the Doppler effect. Stars coming towards us showed a blue shift, stars that were receding a red shift. Hubble found that in distant galaxies there was a systematic trend towards a red shift and for the most distant galaxies the red shift was proportional to the distance, a discovery now known as Hubble's Law. Distant galaxies were receding from us with a velocity that increased linearly with distance. The universe, far from being static, was evolving. It was as if the universe had begun in a colossal explosion, hurling matter out, the fastest bits travelling the furthest.

The Big Bang Story

The standard cosmological model of this evolving universe was nicknamed (scathingly) the Big Bang by Fred Hoyle, whose own commitment was to a Steady-State model of the universe. An account of the Big Bang is the topic of this chapter where we will see that it has had to overcome difficult problems and it has had to cope with many distinguished critics. Is the Big Bang more myth than science? Nobel prizewinner Hannes Alfen, Professor of Plasma Physics in Stockholm, would answer in the affirmative if his article in the *Times Higher Educational Supplement* of September 1976 is anything to go by.

> The big bang is a myth, a wonderful myth maybe, which deserves a place of honour in the colombarian which already contains the Indian myth of a cyclic universe, the Chinese cosmic egg, the Biblical myth of creation in six days, the Ptolomaic cosmological myth, and many others. It will always be admired for its beauty and it will always have a number of believers, just as the millennia of old myths.

Nevertheless, most cosmologists today regard the Big Bang model the best and most plausible. We need to see what it is about and whether it can answer the question, 'Why is anything here?'

From the Solar System to the Milky Way and beyond to other galaxies and galactic clusters, the all-permeating force is the force of gravity. It was natural, therefore, for Einstein to see if the General Theory of Relativity could account for, or at least describe quantitatively, the large-scale features of the universe. There are formidable mathematical difficulties. The curvature of space-time is determined by the distribution of matter and its motion, by radiation, whose energy has gravitational effects, and by the gravitational field itself. Unlike other equations in classical physics, the Einstein Field Equation is a non-linear equation, which means it is not possible to superimpose the effects of separate pieces of matter. Besides which, the gravitational field is an energy field and is therefore, itself, a source of gravity. Mathematical difficulties are not the only problem: there are conceptual problems too relating to the uniqueness of the universe and to any comprehensive treatment of the universe. We have a theory of electrons because there many electrons, all of them identical; we have a theory of the planets because there are more than one and there are common features; we have a theory of stars for the same reason. But there is only one universe. Physics traditionally focuses on a thing and excludes the influence of the rest of the universe as far as possible. But this analytic approach cannot be done for the universe. The universe simply is. Its features can only be passively observed. They are that they are.

If a theory of the universe is possible, if, in other words, cosmology is possible, something more than a sort of religious passivity has to be generated. The Principle of Sufficient Reason — roughly, that there is a reason for everything — must be the guide. Two simplifications are suggested by observation. One is that on the largest scale probed by our telescopes the universe looks broadly the same in all directions, so the universe on the largest scale can be regarded as homogeneous and isotropic. The second is that our position on the fringe of our Milky Way galaxy is hardly likely to be a special one: the universe must surely look the same for all observers. This assumption is then promoted to the status of a principle, namely the Cosmological Principle. (The proponents of the Steady-State universe extended this principle to one in which it is postulated that the universe looks the same from anywhere and *at any time*, which became known as the Perfect Cosmological Principle.)

We are in immediate difficulty by this seemingly commonsensical assumption because it implies that all observers see the same universe as it is now. The difficulty lies in the word 'now'. What does that mean to an observer in the Andromeda nebula, or to one that is so far away that light takes almost the age of the universe to reach us? In order for the Cosmological Principle to have any meaning at all it is necessary to postulate a common time that all observers agree about. Those assumptions about the universe being homogeneous and isotropic refer only to space coordinates, and this immediately decouples time from the space-time continuum. It seems that we have to resort to a form of Newton's Absolute Time if cosmology is to mean anything. We keep the full panoply of General Relativity for small-scale local features like stars and black holes but resort to old-fashioned Time to make sense of the large-scale features of the universe.

The mention of Newton is propitious. Let us dump General Relativity for the time being and invent Newtonian Cosmology. All we are going to use is the inverse-square law of gravitational attraction. The first thing we notice about gravity is that it attracts rather than repels, so why hasn't all the matter in the universe become clumped together? There are three possible explanations. One is that the universe is too young for that to have happened, which presupposes that the matter was spread out when the universe began. The second is that all the matter is moving fast enough to counteract the pull of gravity rather as a stone flung upwards from the earth's surface moves against the gravity of the earth. The third is that there exists an unknown repulsive force that counteracts gravity. This latter explanation was the one that induced Einstein to introduce the infamous cosmological constant into his field equation. The observation by Hubble that the spectrum of light from distant galaxies was, on average, shifted to the red was evidence, based on the Doppler effect, that the universe was expanding. If matter were gradually coagulating as was surmised in our first explanation the shift would be to the blue, so we can rule that out. Which leaves motion and the unknown force.

As motion is actually observed, the simplest model is to drop the idea of a repulsive force and concentrate on motion. Imagine the universe to be a vast sphere. Suppose at a given time, a body is generated with an outward velocity. There are no lurches sideways, no spin, only pure isotropic expansion outwards from the centre.

Then if the velocity is less than the escape velocity, which is determined by the total mass (just as it is on earth by the earth's mass), the body cannot escape. If it is greater, it will escape. Thus, if we imagine an explosion at t=0 that gives bodies a velocity v, expansion continues indefinitely if v is greater than the escape velocity, but it will be followed by a collapse if v is less than the escape velocity.

This provides a simple intuitive picture of the Big Bang. However, we assumed that the density was constant, but in an expanding universe this will not be true—it will get less and less, but the total mass remains fixed, and this can be taken into account. Although the cosmology of General Relativity shares the idea of a universal time with Newtonian cosmology, there are big differences. It is a non-linear theory, so simple summations don't work. Instead of one potential energy, like Newtonian gravity, it has as many as ten, associated with the components of the space-time metric. Energy and pressure are sources of gravitation as well as matter density. Also, space is curved and stretching in time. Apart from the adoption of a universal time some further simplifications are desirable, such as keeping spherical symmetry and regarding matter as being uniformly spread with a certain density. The results are the Einstein Field Equations that describe the rate of change of radius of the universe and its acceleration, determined by the energy-momentum distribution. These equations differ from the analogous Newtonian equations in containing three new factors, namely, Einstein's cosmological constant (which acts as a repulsion opposing gravity), the pressure of the cosmic gas, and a parameter k. k is a constant that describes the curvature of 3-space such that if space is spherical k = 1, if space is flat k = 0, and if space is saddle-like k = –1. It is usual to assume that the velocity of matter particles is much less than the speed of light, in which case, if matter rather than radiation is dominating the gravitation, the pressure can be neglected. If the cosmological constant is also ignored we get a relatively simple scenario. When k=1, the radius exhibits a cyclical variation with time—expansion followed by crunch. When k=0, the radius, R, increases with time according to $R \propto t^{2/3}$.

When k=-1, R increases with time becoming logarithmic at large R. These three solutions are analogous to the Newtonian solutions regarding escape velocity. However, now the three solutions refer to different curvatures of 3-space—spherical for k=1, Euclidean for k=0 and saddle-like for k=-1. Which solution applies to our universe can only be decided by observation, by measuring the rate of expansion and its change with time i.e., the deceleration.

Both the expansion rate and its deceleration are going to be determined by the average density of matter-energy in the universe. There is a certain critical density that makes space flat (k=0). If the universe has this density the expansion will continue forever, but at a decreasing rate ($R \propto t^{2/3}$). If the density is greater than this critical value, space will be spherical, and sooner or later the expansion will change into a contraction. If the density is less, space will be saddle-like and expansion will go on forever. The quantity that quantifies these scenarios is Ω, the ratio of the density of matter-energy in the universe to the critical density that results in flat space. It is therefore of crucial importance that this density factor be measured. If $\Omega = 1$, space is Euclidean and expansion continues forever.

This factor is notoriously difficult to measure, but certain limits can be put on it. Hubble's measurements of expansion rate and distance gives an age to the universe of between fifteen and thirty billion years. If the density is too low, expansion would have reduced the density to almost zero, which is not the case. If the density is too high, crunch would have followed expansion, and there is no evidence of that. Estimates then give limits on the density factor, which must lie between 0.1 and 2. Estimates of the density must come from the observation of visible matter and, if not visible, from its gravitational effects. The trouble is that only a tiny fraction of the universe is luminous. This would be sufficient to make

But there will be an abundance of very faint stars and invisible matter associated with planets and matter dispersed through space. Observation of the orbital rate of stars around galactic centres is explicable only by assuming the presence of a large amount of dark matter in each galaxy that exerts a significant gravitational force. Elementary Newtonian mechanics predicts that stars on the fringes of a galaxy should rotate around the central mass of the galaxy with a period that increases with distance from the centre; just as in the case for planets in our solar system. Observation does not support this prediction: beyond a certain radius the period becomes approximately constant. This can be explained on the basis of elementary Newtonian mechanics only if the galaxy is surrounded by a halo of matter. It is the straightforward gravitational influence of this halo, though invisible, that accounts for the greater than expected rotational speed. Unless Newtonian gravitational theory fails in the limit of weak gravity — and far from the galactic centre the gravity will be weak — the case for the existence of dark matter is strong. All of this can boost the estimate of the density factor to between 0.1 and 0.3. Nobody knows what this dark matter is — some of it may be low-mass stars or super-massive black holes (stars that trap light), or a hitherto undiscovered form of non-baryonic particle (sometimes called WIMPs — Weakly Interacting Massive Particles).

A more direct measure of the density factor is to measure the expansion rate and its variation with time since this gives us the deceleration of the expansion, which can be directly related to the density. Such measurements can be carried out using standard techniques, but they are very difficult in practice. The current results of such measurements are doubly astonishing! For the first seven billion years the expected deceleration is observed, but its magnitude suggests that the density factor should be near unity and space is flat. If this is the case there is a contribution to the density factor of at least

0.7 that is entirely unaccounted for. It is referred to as dark energy. The second astonishing result is that, as the universe has evolved, deceleration has become an acceleration. This implies that some or all of the dark energy gives rise to a repulsion force that now neutralises the effect of gravity. The explanation, such as it might be, can be found associated with the cosmological constant that Einstein originally introduced into his field equations to produce a steady-state universe. The cosmological constant represents a force of repulsion between one piece of matter and another, too small to be so far detected, but, being a constant and independent of expansion, would come into its own once expansion weakened the attractive gravitational force. But what dark energy is remains a mystery.

The introduction of these mysterious features — dark matter and dark energy — is enough to justify Alfen's wry description of the Big Bang as myth. What seems to be the fact that the unknown and mysterious is to be known only in terms of the unknown and mysterious begins to sound like a principle of Hermetic magic. Yet there are features of the Big Bang that are impressively persuasive.

The theory asserts that the universe, space and time included, began some fifteen billion years ago in the form of a stupendously dense dot of matter-energy characterized by a virtually infinite temperature, a dot that became bigger and less dense and cooler with time. Once the fireball had cooled sufficiently, matter recognizable to our earth-bound science appeared. The study of atomic spectra originating in the stars indicates the relative abundance of the elements in the universe. There are good reasons founded on nuclear physics and astrophysics that, beyond the lighter elements like lithium, all the heavy elements are created inside the interior of large stars. When massive stars use up their nuclear fuel — hydrogen, then helium, then carbon etc., — they suffer gravitational collapse and explode, and are observed as supernovae. Their elements are scattered into space, some of which form planets and life-forms. But where did the lighter elements—hydrogen, helium, lithium—come from in the first place? Why are the elements hydrogen and helium so abundant throughout the universe? The Big Bang answer is that these lighter elements were formed when the universe had cooled sufficiently for familiar matter—electrons, protons and neutrons—to appear. The proton is already the hydrogen nucleus. Capture of a neutron creates a deuterium (heavy hydrogen) nucleus and the amalgamation of two deuterium nuclei produces helium. A few other light elements would be produced but the increasing repulsion between a proton and a proton-rich nucleus would effectively eliminate the possibility of forming heavier elements. The more protons and neutrons that were required to form stable nuclei the less probable that those heavier nuclei were formed. The huge energies required to form the heavier elements could only be found in the interiors of stars. Nuclear physics predicts that the abundance of helium relative to hydrogen should be near to twenty-five per cent, and that, indeed, is what is observed in the composition of stars.

In 1965 Arno Penzias and Robert Wilson, worried about the microwave noise being picked up by their antenna at the Bell Telephone Laboratories, discovered the cosmological microwave background. In no matter what direction one 'looked' there was a signal of microwave radiation underlying stronger local sources. Examination of its spectrum showed it to be exactly what is expected of a blackbody at a temperature of about 2.7K, that is, about -270 degrees centigrade. Where did this radiation come from? Once again, the Big Bang Theory had an answer. After the nuclei of the lighter elements had formed the universe would have consisted of a gas of nuclei, mostly protons, and electrons. With further cooling, the protons and other nuclei could capture electrons and the energy released by the captured electrons would appear as high-frequency electromagnetic radiation. This radiation would have a frequency spectrum characteristic of the ambient temperature before becoming largely decoupled from the newly-formed atoms and matter in general, and would then spread far and wide. As time went by the expansion of the universe would produce a red shift in the spectrum of this radiation equivalent to a fall in temperature. The cosmic microwave background of today is the relic of that enormous release of radiation in the early stages of the universe.

So far, so good. But there is a horrendous problem. The cosmic microwave background is too uniform. If its spectrum and intensity are measured in diametrically opposite directions, very nearly the same result is obtained. Opposite regions of the universe have the same radiation temperature which means that they were in thermodynamic equilibrium with each other when radiation was emitted. Thermodynamic equilibrium in a gas can be established only if there is abundant and frequent interaction between the particles that make up the gas. Relativity tells us that there can be no interaction that travels faster than light. If the gas is expanding and the volume of the gas is too large for light to travel from one side to the other within an energy-relaxation time associated with the particle collisions, temperature differences can occur. There is a boundary beyond which temperature evolution proceeds independently. No such boundary seems to exist for the cosmological microwave radiation in spite of expansion, and this requires an explanation. The explanation may be simply that cooling occurred at a rate that was locally the same everywhere so that when the cosmic background radiation was emitted it was always emitted at the same temperature, but it is thought by cosmologists that this is extremely unlikely. If so the fact that opposite regions of the universe exhibit the same radiation temperature is known as the *horizon problem*.

And then there is the *flatness problem*. The evidence concerning the density factor is that it is close to unity, which means that today's universe is flat. Expansion over fifteen billion years reduces the density considerably, consequently if the early density had not been extremely close to the critical density for flatness, the universe today would be far from flat. So how was it that the early universe was flat? How was it that the early universe had that particular special geometry? Well, that was how it was created. Full stop! Ah, but ... there must be a reason. The flatness problem is real. And the explanation is found in the phenomenon of inflation.

Inflation is the brainchild of Alan Guth and others. Shortly after the universe was born there was a brief phase of immensely huge expansion. Elements that were initially close together, and therefore shared the same temperature, suddenly found themselves permanently separated, which solves the horizon problem at a stroke. Moreover, just as a balloon when blown up has an approximately flat surface, so whatever was the initial geometry, after inflation the geometry was bound to be found to be Euclidean. What caused the inflation and what brought it to an end is an ongoing topic of an esoteric theory that involves the concepts of a false vacuum and of negative pressure. But the claim of inflation theory, that it solves both the horizon problem and the flatness problem, is undeniable. Unfortunately, such an immensely huge expansion might be expected to produce intense gravity waves, but these have not been detected. Another observation that may scupper inflation is to do with the tiny temperature variations that are detectable in the cosmic background radiation. These are positively necessary to account for the fact that a non-uniform distribution of matter is absolutely essential to account for the formation of individual galaxies and not one huge clump of matter. A little excess of matter in one place will attract nearby matter and so will grow, and a little excess of matter there will do the same. Hence the rich population of individual galaxies that is the principal feature of our universe. Inflation theory is quite happy that there should be quantum fluctuations in the background radiation, but the statistics are expected to show that the fluctuations are random. There are some recent observations that this is not the case. If this is confirmed, inflation theory will have another problem to solve.

In fact, inflation theory has become somewhat notorious in suffering problems, so much so that it has stimulated alternative scenarios, inevitably labelled heresies. One of these scenarios questions the received wisdom that the aether, the medium that classical physics had postulated as necessary for the propagation of light, does not exist. Although experiments designed to detect the aether gave negative results, there are arguments that claim that this did not necessarily rule out its existence. If the aether existed as a gas of exceedingly tiny particles that supported wave motion like any other gas, then the speed of light would depend on the temperature. If so, it would imply that the velocity of light in the early universe, where the temperature according to Big Bang theory was enormous, would be much greater than it is today. Consequently, thermal communication between distant parts of the universe would be extremely rapid and efficient, which would explain why the temperature of the cosmic background radiation is so uniform. In other words, the assumption of a superluminary speed of light in the early universe would solve the horizon problem without the need for inflation.

But if the velocity of light is no longer the fundamental constant we assume it to be, what of the others — the gravitational constant, Planck's constant, the charge on the electron? Standard cosmological theory assumes that they have not changed over the life of the universe. But perhaps most serious of all pertaining to any account of the very early universe is the lack of a quantum theory of gravity. Quantum gravity in short. Until that is in place it is difficult to avoid Alfen's imputation of myth.

Of course, the Big Bang theory has had its critics from the outset. As we have already remarked, the very term Big Bang was coined by Fred Hoyle, one of its staunchest critics, who remained so to his recent death. The central idea supporting the early criticism was the idea of a Steady-State universe. It was encapsulated by the Perfect Cosmological Principle which postulated that the universe should not only appear the same whatever the position of the observer but also it should appear the same at all times. In the model, the observed fact of expansion was countered by the continuous creation of matter, an idea that, surely, Henri Bergson would have approved. Creation of matter there had to be whatever the cosmological model. In the Big Bang theory the matter was created once and for all, in the Steady-State theory the process was continuous. The discovery of the cosmic microwave background and its plausible interpretation in the Big Bang convinced most, and still convinces most, that the Big Bang theory was correct and the Steady-State theory was wrong.

But recently the Steady-State theory has been brought up to date by Hoyle, Burbidge and Narlikar as the Quasi-Steady-State theory. Perhaps encouraged by the imaginative postulation of unobserved fields elsewhere in physics, they revive the idea of a matter-creating field that, like the false vacuum of inflation theory, exhibits negative pressure. A creation event is assumed to occur and induce an expansion. The universe reaches a maximum radius and begins to contract and reaches a minimum again at which point there is another creation event. The solution is therefore oscillatory and continues forever. In the developed Quasi-Steady-State model the universe oscillates with exponentially increasing amplitude.

The most attractive feature of the model is that expansion and the creation of matter arise from the same field, and overall energy is conserved. There is also no singularity as there is in the Big Bang theory and no need for inflation since, it is claimed, that thermalization can proceed over many cycles. The most difficult features are the model's account of the uniform temperature of the cosmic microwave background and its explanation of the helium/hydrogen ratio in the absence of a primeval fireball. But the Big Bang theory also has its problems as well as its undoubted successes. It is disturbing to some degree that both theories postulate fields that have yet to be observed. The Big Bang has the most adherents today, partly, one suspects, because of the well-publicised failure of the early Steady-State theory. It is to be hoped for the health of cosmology that enthusiasm for alternative theories continues to escape the charge of heresy, which is a hope for the reformation of science.

Magnificent though these cosmological theories are, and monuments to the Principle of Sufficient Reason, their dependence on abstruse mathematics that allows time loops and singularities, and the advancement of hypothetical models that seem far from the possibility of empirical verification or falsification, give rise to the feeling, perhaps understandably, that what is presented in modern cosmology is often considered to be more myth than science. Moreover, the tendency of its practitioners to be dismissive of alternate theories has more than a whiff of mathematical theology rather than rational exploration, in a way that reminds us of fundamentalist Darwinists. Yet, unless the interpretation based on General Relativity is totally wrong, the universe *is* expanding; moreover, the cosmic microwave background *does* exist, and the helium to hydrogen ration *is about* twenty-five per cent as expected from extrapolations to high temperatures of the results of terrestrial measurements in nuclear physics. The Big Bang is the simplest model to explain these observations. Maybe the problems that the theory runs into—the horizon problem and the flatness problem, for example, are because of the attempts to push a description back to an imagined beginning, and these can be solved by a reappraisal of the starting conditions; starting conditions, for instance, that postulated flatness and thermodynamic equilibrium in a more modest universe that was accessible to terrestrial physics. Admittedly, limiting the proto-temperature, say, to that required for helium formation, would make cosmology less interesting to high-energy particle physicists, and much of the myth-like nature would be lost. If such a cosmology were feasible, we would have a new Big Bang model, the Prudent Big Bang model. It would be less exciting, but it would be less recognizably pure myth.

Chapter Thirteen

Meta-Cosmology

Prudence is unexciting. The modern trend in physics is towards the opposite, as a brief look will indicate. What could be more imprudent than meta-cosmology, the study of multiple universes. In one sense, the concept of many universes has the venerable authority of the philosophy of Plato in the form of the Principle of Plenitude — if a thing can exist, it will. If thinking can make it so, there is no shortage of alternative universes. Leibnitz was the first to conclude that of all universes we live in the best of all possible worlds. What does science say about that, if anything?

One of the first many-universe theories emerged out of quantum theory, following the logic of the superposition principle of quantum theory. It was to furnish the universe with a wave-function as a necessary step towards a quantum cosmology. Consider the famous two-slit experiment exhibiting the wave-particle nature of the electron. As a wave, the electron goes through both slits and is detected on the measuring screen as a particle. The transition from wave to particle is commonly referred to as the collapse of the wave-function. Such a collapse is not predicted by Schrödinger's equation. What is predicted is the superposition of the electron wave with the wave-function of the measuring screen and the subsequent evolution thereof, ad infinitum. According to the Many Worlds theory of Hugh Everett III, the screen splits into all the possibilities of the outcome of the measurement. We observe only one of these possibilities, but, in fact, the wave-function does not collapse; it continues to evolve in Many Worlds. Our world is just one of many. Moreover, the wave-function of the universe can be understood only in the context of a Many Worlds cosmology.

All of this follows inexorably from quantum theory as we know it. Perhaps it is incomplete as Einstein thought. Certainly, no one has yet found a way of including gravity, which is, at least at the cosmological level, a monumental incompleteness. In view of this, an overdue application of Occam's razor might be in order and evidence of quantum humility regarding the cosmological status of Schrödinger's equation could be appropriate.

A more interesting idea came from Brandon Carter in 1974 who invented the Anthropic Cosmological Principle which, to some degree, put Man back in the centre of things. He advanced the hypothesis that there existed numberless co-existing universes, each of them logically consistent and compatible with Big Bang theory. The motivation was to understand why our universe was so well-matched to the large-scale needs of humanity. The evolution of our universe was just right and old enough for matter to cluster into galaxies, for stars to form and manufacture the elements that compose organic life, and for there to be time for supernovae to occur and scatter those elements everywhere. The Anthropic Principle affirms that the universe had to have just those properties it is observed to possess for man to appear.

The fact that there were Laws of Nature and innumerable subtleties of detail that allowed mankind to exist was an argument for the existence of God from design. What were the chances of the universe having just the right properties? Surely it was proof that an all-wise being created the world: God existed. Not so, claimed Carter. If a meta-universe existed that consisted of all possible sorts of co-existing universes, then it is no surprise to find that we live in one that suits us, one that simply came about by chance. No need to evoke a God, at least, not via the argument from design.

One can envisage a sequence of universes in time. Accepting that Big Bang theory could easily give rise to universes very different from the one we inhabit, the late John Wheeler suggested that if there were many universes, one following the other, a suitable universe was bound to appear sooner or later. In this sequential cosmology, a universe is born without memory of its predecessor, with properties governed by the Laws of Chance. Sooner or later our existing universe was bound to happen. Once again, no need for God. But this appeal to the Laws of Chance makes the argument rather weak. Given the complexity of our universe, or indeed of any hugely complex gambling system, there is absolutely no guarantee that a particular permutation will ever turn up. As Ian Hacking has pointed out, merely waiting long enough will not do. Moreover, having got the goods is certainly no evidence that therefore there have been many attempts before. We may have just got extremely lucky first time around. The plenitude cosmology does not have this problem — all universes are possible, by hypothesis, so ours already exists.

That the world was designed purposely for us by God goes back as we have seen to the teleological argument of William Paley in 1802. The weapon that cosmologists have used to counter this idea is statistics — it all could have happened by a combination of chance and necessity. By necessity is meant the Laws of Physics as applied, insofar as they can be to Big Bang theory. Chance is distinctly more problematical. Does it apply to the Laws of Physics themselves? If so, in what sense? Do we envisage a range of magnitudes for Planck's constant, for example, with some statistical distribution? Do we stick to three spatial dimensions, or imagine more dimensions or less, or even fractal dimensions? Why not? Anything goes. The fact is, that bringing chance into the provenance of the universe means that there has to be a lot of them. Nothing scientifically sensible can be said about a unique object. *It is that it is*. Full stop. So another cosmological model has to be one in which there are co-existing universes, in order, as Stephen Hawking puts it, to provide an ensemble with some statistical distribution. The result is similar to the plenitude cosmology but without the Platonic claim.

A meta-cosmology of recent origin is another sequential-universe cosmology, but this time with memory, and it has the distinction of being based on a cosmology quite different from the Big Bang theory. In the Quasi-Steady-State model of Hoyle, Burbidge and Narlikar one universe grows out of the death throes of its predecessor; it expands and ultimately contracts to a fire ball, out of which grows a new universe. The authors claim that the uniformity exhibited by our universe is a consequence of there being a long sequence of causally connected universes. There is clear room here for the hypothesis to be made of a steady evolution towards a universe that supports life. I submit, in all modesty, for the reader's attention, the possible existence of a Cosmological Theory of Evolution.

The advantage of having many universes one can play the game of what if? What if the fundamental constants — Planck's, the velocity of light, the gravitational constant, the electron charge, etc. — were different? In most cases it would turn out that we shouldn't be here at all. What if those fundamental constants were not constant at all, but time-dependent? That would certainly change our ideas of the Big Bang. What if the world had a different number of space dimensions? Would we survive? In a meta-universe with an unlimited number of universes, there would be some with exactly these odd characteristics, and several more. So, do we live in the best of all possible worlds? The Anthropic Principle assure us that our universe, if not the best, is good enough to explain why we are here. In their book *The Anthropic Cosmological Principle* (1986) John Barrow and Frank Tipler provide a definition.

> The observed values of all physical and cosmological quantities are not equally probable but they take on values restricted by the requirement that there exist sites where carbon-based life can evolve and by the requirement the Universe be old enough for it to have already done so.

The Principle makes sense only if a meta-universe exists that consists of universes that have fundamental constants chosen at random. But even if the idea of a meta-universe is just a creation of mathematical theology, it serves a purpose outside any mathematical brief. That purpose is a religious one, religious in Whitehead's sense, and the Anthropic Principle gives a focus to that purpose, which is to evoke awe. Without the idea of a meta-universe the Anthropic Principle is as vacuous as that mindless, but deeply metaphysical, chant at a football match: 'We're 'ere 'cos we're 'ere.' Instead, in an entirely secular way, it draws the attention, Paley-like, to the delicate design of the universe and to the fact that we exist because of that delicate design.

Nowhere is the delicacy of design exhibited more strikingly than in the production of elements heavier than helium in stars. George Gamow (1904–68), sometimes referred to as the originator of the Big Bang Theory, was one of the first to investigate the problem of nuclear synthesis with his student Ralph Alpher. Gamow's humour was irrepressible. In a paper submitted to the *Physical Review*, he persuaded Alpher to add the name of the famous astrophysicist Hans Bethe. To his delight, the paper by Alpher, Bethe and Gamow was published on 1 April 1948. In his study of the creation of the elements in the primeval fireball there was a problem. Hydrogen with one proton was stable; deuterium, with two particles, a proton and a neutron, was also reasonably stable and so could be produced from hydrogen by the capture of a neutron; so was tritium with three particles, a proton and two neutrons, and all could be converted into helium, a stable atom with two protons and two neutrons. Further progress by neutron or proton capture was severely inhibited by the absence of any stable nucleus with just five particles. So how could heavier elements be produced? Here is Gamow's version of genesis (from Alan Guth's book *The Inflationary Universe* (1997)).

> In the excitement of counting, He missed calling for mass five and so, naturally, no heavier elements could have been formed. God was very much disappointed, and wanted first to contract the universe again, and start all over from the beginning. But it would be much too simple. Thus, being almighty, God decided to correct His mistake in a most impossible way.
>
> And God said, 'Let there be Hoyle.' And there was Hoyle. And God looked at Hoyle ... And told him to make heavy elements in any way he pleased.
>
> And Hoyle decided to make heavy elements in stars, and to spread them around by supernova explosions.

The crucial step in the formation of the heavier elements in stars is the fusion of three helium nuclei (alpha particles) to form an isotope of carbon plus two gamma rays.

The probability of three alpha particles coinciding is small so this reaction is expected to be too slow. However, the probability of two alpha particles coinciding will be much better, so an intermediate step could be the formation of an isotope of beryllium followed by an absorption of a third alpha particle. This scheme is promising because beryllium has a comparatively long lifetime, but the second step may be too slow. Fred Hoyle conjectured that since carbon undoubtedly exists in the universe, it was likely that there was an excited state in the carbon nucleus whose energy was close to the energy level in the compound nucleus formed by the amalgamation of helium and beryllium, in which case the reaction would go rapidly. This, indeed, was found by nuclear physics to be the case. There was a small difference that could be easily made up by the positive kinetic energy of the carbon nucleus in the hot stellar interior. So Hoyle's conjecture proved correct. However, a further reaction with another alpha particle could destroy the carbon rapidly through the production of oxygen if there was another resonance. It turned out that the relevant level for oxygen was fractionally too high, so no amount of kinetic energy of the oxygen would produce resonance: in fact, it would make the discrepancy worse. Carbon was saved for organic life thanks to the unusually long life of the beryllium isotope, the carbon resonance, and the lack of resonance in oxygen! Looking at the magnitudes involved it is difficult not to think that it could so easily have been different.

Tinkering with the fundamental constants could very easily destroy the carbon resonance and much else. It may be safer to limit attention to the dimensionless quantities that they give rise to. Arthur Eddington (1882–1944), sometimes called the first astrophysicist, was fascinated by the dimensionless numbers that cropped up in physics. There were three that he believed could be calculated on a priori grounds. They were the reciprocal of the fine-structure constant (which determined the strength of interaction between the electron and the photon), the ratio of the mass of the proton to that of the electron, and the cosmical number, the number of protons in the universe. An account of the relevant calculations were published after his death in his book *Fundamental Theory* (1948). The fine structure constant is formed out of the electronic charge, the speed of light and Planck's constant, and its reciprocal is very close to the number 137. Eddington sought (successfully, he believed) to account for this number by a priori argument. In his book *The Philosophy of Physical Science* (1939) he famously began a chapter:

> I believe there are 15, 747, 724, 136, 275, 002, 577, 605, 653, 961, 181, 555, 468, 044, 717, 914, 527, 116, 709, 366, 231, 425, 076, 185, 631, 031, 296 protons in the universe, and the same number of electrons.

Few nowadays take Eddington's idea seriously. But, still, the dimensionless numbers of the universe exert a fascination. Paul Dirac (1902–84), famous for his quantum-field theory of the electron, noted the existence of many physically significant numbers enormously larger than the reciprocal of the fine-structure constant and the proton-electron mass ratio. Eddington's cosmological number was one ($N_1 = 1 \times 10^{79}$), the ratio of the age of the universe to the time it takes light to cross an atom ($N_2 = 6 \times 10^{39}$) was another; a third was the ratio of the electrical to the gravitational force between a proton and an electron ($N_3 = 2 \times 10^{39}$). Within a numerical factor of order unity, we have $N_2 \approx N_3$ and $N^1 \approx N_{2,3}^2$. Dirac believed that these relationships could not have come about by chance, and so he proposed the *Large Numbers Hypothesis* in which he claimed that nature somehow decreed that any two of the large numbers are connected by a simple mathematical relation.

One of the consequences of this was that, because one of the large numbers, N_2, contained the age of the universe, such numbers had to be time-dependent. This meant that in the case of N_3 either the elementary electric charge had to increase with time or the gravitational constant had to decrease with time. The former would have alarming consequences throughout quantum mechanics, so Dirac chose to confine the time-dependence to the gravitational constant, which meant that it became inversely proportional to time, and it follows that the number of particles in the universe increased as the square of time. Dirac's suggestion created a welter of new cosmological theories that incorporated the time variation of the gravitational constant, and which are, in principle, testable.

Generalising the time-dependence to the dimension of the atom led mathematical physicist James Jeans to explain the expansion of the universe in terms of the shrinkage of the atom with time. A smaller atom means that the electrons would be more tightly bound and therefore their binding energies would increase. Emission of radiation would involve shorter wavelengths in smaller atoms. Thus, over time, as atoms shrank, the light they emitted would shift towards the blue. Younger atoms, those in distant galaxies, would emit light which, compared with atoms on earth, would be shifted towards the red. Jean's idea has been criticised in implying the elementary charge, or Planck's constant, or the velocity of light, or any two, or all three, have to be time-dependent, which, if true, would be severely disruptive. Nevertheless, as Hoyle has pointed out, the expansion of the universe assumes that the properties of matter do not change.

We might, finally, wonder whether we could exist in a world with different space dimensions, since in our meta-universe there are one-dimensional, two-dimensional, and n-dimensional universes. It turns out that *only in three dimensions* does light travel with a fixed speed, otherwise some light travelling slower than c (never faster) becomes superimposed on the light travelling with c, and one gets blurring. Also, only in three dimensions can light travel without distortion. So we have the best of all possible dimensions.

The Anthropic Principle introduced by Brandon Carter is unobjectionable, but there are stronger versions, and the original is now known as the *Weak Anthropic Principle*. The *Strong Anthropic Principle* says the universe *must* have the right properties for life. The *Final Anthropic Principle* gives up on life and substitutes intelligent information-processing: 'Intelligent information-processing must come into existence in the Universe, and, once it does, it will never die out.'

This claim is couched in familiar mechanistic language. The human being has become a purely physical object, a biochemical machine functioning according to the laws of physics. It is therefore, not entirely surprising to find that Frank Tipler, one of the authors of *The Anthropic Principle* has written a book with the arresting title *The Physics of Immortality* (1994). He begins:

> This book is a description of the Omega Point Theory, which is a testable physical theory for an omni-

present, omniscient, omnipotent God who will one day in the far future resurrect every single one of us to live forever in an abode which is in all essentials the Judeo-Christian heaven.

The time has come, it appears, to absorb theology into physics. As a physicalist, Tipler makes the valid point that either theology is irrelevant, or it becomes a part of physics. Where most scientists, being atheistically inclined, might opt for the irrelevancy, Tipler makes a remarkable case for, what is essentially, the physics of God. But you have to buy in to physicalism. Never mind.

Chapter Fourteen

Beyond Belief

In this book I have tried to emphasise the gap between science as popularly exposed and experienced and the intellectual and spiritual yearning for purpose and meaning that each of us has. In many ways, science has become the new Church, with its various dogmas and its all-pervading materialist message. It certainly has its answers to the questions of Theodotus:

> who we were, and what we have become; where we were...whither we are hastening; from what are we being released; what birth is, and what is rebirth?

but they are inevitably technical and somewhat miss the point. It responds magnificently to the material needs of mankind, but ignores completely that religious impulse eloquently described by Alfred North Whitehead (see p. 14 above). Einstein summarizes the situation succinctly:

> Science without religion is lame, religion without science is blind.

Contemporary science is flawed on two counts: it manifests a dangerous trend towards absolute dogmatism as regards its theories, and it scorns the concepts of mind and soul and the religious impulse on man. It is in need of reform.

There is nothing like science for producing turmoil. After evolution, look at IQ. Testing for something like intelligence is bound to be fairly dodgy, yet the scientific measurement of IQ, the intelligence quotient, has turned out to be as good a predictor of human performance as any in psychology. It measures something that is real, a sort of general mental ability, and this is where the tumoil comes in, it seems to be in large part hereditary. Believers in nurture rather than nature can't go along with this, and it is not surprising that IQ and books like The Bell Curve become politically incorrect in socialist and tender-minded circles. Even more politically incorrect is its racist results. Applied globally, IQ testing appears to prove that Orientals are more intelligent than Europeans. Why passionately dismiss it, given the extensive experimental evidence? As elsewhere, it is a case of being beyond belief. And the dogma, in this case is all on one side.

Regarding the censorial role of establishment dogma, perhaps there is no better example than that provided by the tale of the black hole in astrophysics. Received wisdom has it that there are stars that have become so massive and dense that they have become black holes, objects that the general theory of relativity tells us that light, fast though it is, cannot escape from its vicinity. Newtonian gravitation theory says something similar as regards bodies with mass. If the star is sufficiently massive such a body trying to escape the gravitational field would need to have a velocity greater than the speed of light, which according to relativity is impossible. Light does not have a mass, so Newtonian theory does not apply. In general relativity, gravitation induces a curvature in space, a change in the surrounding geometry, that serves to trap light. Nobody as yet, it seems, has positively identified a black hole among the stars—it would have to be through its gravitational effect on its neighbours—but who doubts their existence, that they are engines for the active galactic nebulae?

Well, it turns out that doubters do exist, though you would not know it from a perusal of the popular and semi-popular literature. Given the vast intellectual input there has been on their properties, it would be sad if their existence proved to be merely a misinterpretation of an element of Einstein's theory. Yet a misinterpretation is what those heretics claim. The trouble here is, so mathematically esoteric is the theory, that it is difficult for those of us outside the field to form an opinion as to the correctness of the claim. But given that the claim is advanced by reputable theoreticians, one can, at least, have the opinion that the problem should be aired more widely than it is.

The idea of a black hole originated in the description of the space-time surrounding a static spherical mass of arbitrary gravitational strength by Karl Schwartzchild in 1916. He discovered a mathematical singularity—a term becoming infinite—that occurred at a certain radius from the centre of the spherical mass, which has become famously known as the Schwartzchild radius. At this radius the velocity of light vanishes, and this was widely interpreted as the signature of a black hole, though Schwartzchild did not call it such. The question appears to turn on the interpretation of the radius. Critics claim that this radius does not refer to a real distance from the centre of mass, but is only a distance inferred by an observer situated infinitely far away, and therefore residing in an altogether different space. Referring everything to an observer situated in the Schwartzchild space eliminates the singularity. In other words, it is claimed that if a coordinate transformation eliminates the singularity, it cannot have any physical reality, and therefore the black hole does not exist. Since the point bears on the astrophysics of gravitational collapse, it is strange that seemingly no awareness of this state of affairs exists. It supports the view expressed throughout the book that a reformation of science is overdue.

Black holes apart, science has been endlessly blamed for the decline of religion in the West. That is, perhaps, one reason why the muslim world has virtually turned its back on science. But I wish to distinguish scientismus from pure science, and to defend pure science wholeheartedly. Doubtless, faith in scientific truth is stronger than faith in religious truth, insofar as scientific truth is demonstrable. Indeed, as far as scientific truth is concerned, it would be better to say, pace some strands of modern philosophy, that faith hardly comes into it. Faith, it may be argued, is needed only when there is doubt, and who can doubt the success of science? Even so, there is no logical reason why a belief in science should drive out a belief in God. After all, many scientists believe in God. Of course, science says nothing about God, nor can it: the religious world and the world of science exist in parallel. That the Law of Gravity tells us how the heavens move; that the Laws of Quantum Electromagnetism explain the transistor and the laser; that the Laws of Nuclear Physics gives us unlimited power; none of this can possibly diminish the certainty that a complementary world exists that is inhabited by morality, art, love, passion and meaning. The evidence is the existence of our own conscious minds. Any scientific belief that the one real world is the material world revealed by science is not the fault of science, it is the fault of a failure of the imagination, of intuition, scientismus, in a word. Pure science cannot be blamed for that. Indeed, it is more likely to stimulate the imagination through its marvellous insights into the complexity of Nature, than to impose a sort of materialistic straitjacket. Just by labelling something gravity, or an electron, or a photon, does not eliminate the mystery of Nature, the labels merely facilitate our manipulation of natural things. Nature will always remain deeply mysterious, even in its purely material guise. All the more reason for not discounting the religious impulse.

One reason for the success of science, which at the same time has had a baleful effect on religion, is the method of reductionism, matter being reduced to its elementary parts. It works productively everywhere outside of life. The body is reduced to its interacting organs, the organs to cells, the cell to its constituent molecules, the molecule to its atoms, the atom to its electron orbiting the nucleus, the nucleus to its protons and neutrons, and these, in turn, to quarks and gluons. The reductionist message is that we are nothing more, at base, than quarks and gluons, that we are nothing more than matter. All of which is fine. The baleful effect is in the deduction that therefore there can be no such thing as the world of the spirit. There are at least two objections to this materialist deduction. The most basic is that it rests on a prejudiced view of matter; matter is that, the behaviour of which, is exhaustively describable by science. A second objection is derivable from science itself. As a physicist who has been fascinated by the electronic properties of semiconductors for many years, I have got along quite nicely without bringing quarks and gluons into any theory. Certainly, semiconductors can be reduced to those elementary occupants of the Standard Model, but it would be unhelpful. The interesting properties of semiconductors stem from atoms being arranged in a crystal lattice. The point here is that crystal physics generates concepts and theories that are inapplicable to atomic, nuclear or particle physics. The crystal is an emergent manifestation of matter, and introduces properties that atoms, and nuclei do not have. The whole is more than its parts, in other words. As matter becomes organized in more complex ways, as occurs in living bodies, new properties appear that are not directly related to its reductionist base; consciousness, and indeed life, are two such emergent properties.

An even more baleful influence than reductionism was determinism, and, this time, science must accept some of the blame. The Laws of Mechanics were held to determine absolutely the future behaviour of any system once the initial state of the system had been specified. Applied universally, it killed any idea of freedom of choice; good acts and bad acts were predetermined; there was no meaning to morality and hence no point in religion. In a rationalist furore, determinism was triumphantly claimed to be the inescapable fact revealed by physics. It was an entirely unjustified claim. The fact of the matter is that the two-body problem in mechanics is the only problem in classical physics that can be solved accurately. The three-body, four-body, n-body, problems can be solved only approximately. In the study of gases it is impossible to specify the exact starting conditions of the motions of millions and millions of atoms or molecules, and, even if these were known, it would be impossible to produce an accurate solution. Thus, at best, there was only ever a kind of statistical determinism. More recent studies of non-linear systems reveal that any accurate deterministic prediction would require a knowledge of the starting conditions of impossible precision. Finally, the discovery of quantum phenomenon meant that strict determinism was actually ruled out in nature, only probable behaviour allowed. What this means is that determinism is no longer the bane of freedom of choice. Freedom of choice may be only approximate, but it is there.

The importance of religion in the West may have declined as a result, however illogical, of the success of science, but it is still very much in evidence, especially in the States. Churches there remain full and creationists abound. Much less so in the UK, where the replacement, if it can be called a replacement, is Humanism. Belief in the supernatural, or at least a fascination with it, is still widespread, witness the literary success of Harry Potter, Pulman's *His Dark Materials* and Tolkein's *the Fellowship of the Ring*. We do like myths. Not only like them, but seem to need them.

Nowhere has this been more evident than in the global warming controversy. On the one hand, the scientific measurements of the carbon dioxide content of the atmosphere, which no one disputes, shows that it has risen by some 30% since the beginning of the Industrial Revolution. It was claimed that, since carbon dioxide (CO_2) is a greenhouse gas, one that inhibits the escape of infrared radiation from the surface of the earth, the increase in its concentration would cause a global warming which would catastrophically alter the world's climate, causing the polar ice caps to melt and increase sea levels drastically. Computer simulations supported these predictions. Beyond a certain concentration—the tipping point—it was further claimed that global warming becomes irreversible, and the earth, like Venus, would become uninhabitable. This apocalyptic view was supported by the Royal Society, the American Physical Society and other Western scientific societies (but not the Russian). It was endorsed by the political establishments of the USA and Europe, who, convinced that the scientific case was established beyond reasonable doubt, became busy making laws designed to reduce further emissions of CO_2 and attempting to persuade the burgeoning economies of India and China to follow suit.

On the other hand, there existed a body of globally reputable climatologists who argued the scientific case, far from being established, was fundamentally flawed in focusing exclusively on the effect of CO_2 in the atmosphere. They pointed out that the powerful roles of fluctuations in the supply of solar radiation, of cosmic rays, of water vapour and clouds, were not taken into account in the computer models. In short, their criticism was that the forces that determine climate change were extremely complex and poorly understood, and to claim that the cause of global warming was known with certainty was not recognizable scientific practice. Worse, given evidence of global cooling in recent decades (refuting all computer predictions), the claim of certainty began to look more a product of religious fanaticism than of science, a view that carried more and more conviction with every utterance of the Greens and other environmentalist organizations. But, as Darwin's bulldog Thomas Huxley warned: 'Belief, in the scientific sense of the word, is a serious matter and needs strong foundation.' On any dispassionate view, it was argued, a strong foundation for the claim that global warming existed, and that it was caused by the man-made emissions of CO_2, simply did not exist. It followed that any belief in global warming, if entertained at all, ought to be severely provisional.

Whatever the truth of it all, science does not come out of this very well. For a start, scientific societies, established to serve science, have no business making ex cathedra statements on scientific issues. Improving natural knowledge is the business of the individual scientist talking to his peers. It was not the Royal Society that discovered gravity, it was one Isaac Newton. Furthermore, for scientists to persuade politicians to pontificate on scientific matters is to devalue and endanger science itself. But scientists are, after all, human, and enjoy medals and knighthoods. No doubt scientists "only being human" account for the obscenities of state-run science in Soviet Russia and Nazi Germany. There have also been claims that some scientists have tinkered with the facts to promote their case for more funding. It is the usurping of science by its Societies, the mixing of science with politics, the suggestion of scientific fraud, that are among the most disturbing elements of the global warming controversy. All of which underlines the need for a scientific reformation. Especially if, after all, global warming, so passionately accepted by a distinguished set of the Western scientific establishment, turns out to be a myth.

Of the topics treated in this book, three might be further candidates for serious modern myths. One is Bergson's account of creative evolution in terms of the élan vital; another is cosmology's vast canvas that includes the Big Bang and meta-universes; the third is the esoteric movement of particle physics towards *The Theory of Everything*. Can I hear the reader say, 'With myths such as these, who needs institutional religion?' Richard Dawkins claimed that his book *The Blind Watchmaker* was written:

> in the conviction that our own existence [was]once presented as the greatest of all mysteries, but that is a mystery no longer because it is solved.

On the other hand, as one of the characters remark in *Who's Body* by Dorothy L. Sayers:

> 'There's nothing you can't prove if your outlook is only sufficiently limited.'

I doubt whether David Hume would have been convinced that that particular mystery had been abolished by the theory of evolution. In *The Natural History of Religion* he desperately advocates philosophy.

> The whole is a riddle, an enigma, an inexplicable mystery. Doubt, uncertainty, suspense of judgement appear the only result of our most accurate scrutiny, concerning the subject. But such is the frailty of human reason, and such the irresistible contagion of opinion, that even this deliberate doubt could scarcely be upheld; did we not enlarge our view, and opposing one species of superstition to another, set them a-quarrelling; while we ourselves, during their fury and contention, happily make our escape into the calm, though obscure, regions of philosophy.

As one wit has it:

> Philosophy is questions that cannot be answered. Religion is answers that cannot be questioned.

In view of all of this, maybe we had better get on with our lives and live. There are no solutions. To be without hope is to be happy. No doubt every now and then, that quantum of reason, as Nietzsche puts it in *The Twilight of the Gods*, will fade and all the power of that cosmic religious awe will envelope us once more.

> If one spends oneself on power, grand politics, economic affairs, world commerce, parliamentary institutions, military interests—if one expends in this direction the quantum of reason, seriousness, will, self-overcoming that one is, then there will be a shortage in the other direction.

We are once again in the throes of that General Principle of Complementarity, exuberant life driving out cosmic wonder, cosmic wonder driving out exuberant life.

What can be retrieved? Well, for certain, I have a soul, if what is meant by soul is the form of my mind and body, and presumably, my unique DNA. But is it immortal, as so many great minds have claimed? Sorry, I can't go along with that. In any case, what does it mean to claim a soul for a colony of cells? I am what we are! If there is one attribute of life that is clearly evident, it is that nothing living persists for long. To put it in Bergsonian terms, life is a continual becoming and not as any form of being, including an immortal soul. The idea of immortality is anyway somewhat distasteful, an insult to life, useful to society only as a threat of judgement after death guaranteeing morality. As for that, there is really no need for it; the qualities of the human mind are sufficient. The wonder of consciousness, there for all to experience directly, is the fundamental phenomenon. It embraces ethics, aesthetics, mathematics and science, equally at home with the *a priori* and the *a posteriori*, the defining arche of humanity. It is the existence of consciousness, rather than the myth of immortality, that must engage our attention.

It amounts to this: one must choose one's beliefs with care, and none should be held other than provisionally. One cannot live without beliefs, but they must serve, not dominate. In a real sense, we must be beyond belief. The world has seen enough of fanatics—religious, racist, heralds of the apocalypse, animal rights, global warming—the list is endless. The fear of death, the awareness of man's vulnerability in an indifferent cosmos, the sense of pessimistic inferiority, all these negative sentiments serve to breed the sort of beliefs that must not be questioned. True religion, defined by the glorification and confident delight in exuberant life, is thereby buried under obscene dogma. It is utterly grotesque that the same negative sentiments tend to inform some of today's science. Science must be, simply, beyond belief.

Sources of Quotations

Chapter 1
p. 13 *Science and the Modern World* (Penguin Books, 1938)

Chapter 3
p. 35 Comment in *Somnium Scipionis I* (Frances Yates, *Giordano Bruno and the Hermetic Tradition*, University of Chicago Press, 1964)
p. 37 *Corpus Hermeticum XI* (Ibid.)
p. 38 Ibid.

Chapter 4
p. 54 *Essai philosophique sur les probabilités* (1814)
p. 55 Quoted by Bergson in *Creative Evolution* (1911)

Chapter 5
p. 59 'On Academies' in *Letters on the English* (Harvard University Press, 1919)

Chapter 6
pp. 77–8 *The Principles of Nature and Grace, Based on Reason* (1714); *Refutation of Spinoza* (1708).
p. 80 Byron (1788–1824) *Don Juan* c. xi. i
p. 80 *Boswell's Life of Johnson*

Chapter 8
p. 102 *Contemporary Materialism* (ed. Moser and Trout, Routledge)

Chapter 11
p. 152 *Creative Evolution* (1911)

Short Bibliography

A History of God by Karen Armstrong
   (Ballentine Books, New York 1993).
Darwinian Fairy Tales by David Stove (Avebury, Aldershot 1995).
Exploding a Myth by Jeremy Dunning-Davies
   (Horwood, Chichester 2007).
God and the New Physics by Paul Davies
   (Simon and Schuster, New York 1992).
Heaven and Earth by Ian Plimer
   (Taylor Trade Publishing, Lanham Maryland 2009).
History of Western Philosophy by Bertrand Russell
   (George Allen & Unwin Ltd., London 1961).
Life Ascending by Nick Lane (Profile Books, London 2009).
Science in History; Volume 2 The Scientific and Industrial Revolutions
   by J. D. Bernal (Penguin Books, London 1969).
Science, A History by John Gribbin (Penguin Books, London 2002).
The Fabric of the Cosmos by Brian Greene
   (Penguin Books, London 2004).
The God Delusion by Richard Dawkins
   (Houghton Miflin Company, Boston 2006).
The Real Global Warming Disaster by Christopher Booker
   (Continuum, London 2009).
The Trouble with Physics by Lee Smolin
   (Penguin Books, London 2007).
The Undergrowth of Science by Walter Gratzer
   (Oxford University Press, 2000).

Short Bibliography

A History of God by Karen Armstrong
  (Ballentine Books, New York 1993).
Darwinian Fairy Tales by David Stove (Avebury, Aldershot 1995).
Exploding a Myth by Jeremy Dunning-Davies
  (Horwood, Chichester 2007).
God and the New Physics by Paul Davies
  (Simon and Schuster, New York 1992).
Heaven and Earth by Ian Plimer
  (Taylor Trade Publishing, Lanham Maryland 2009).
History of Western Philosophy by Bertrand Russell
  (George Allen & Unwin Ltd., London 1961).
Life Ascending by Nick Lane (Profile Books, London 2009).
Science in History; Volume 2 The Scientific and Industrial Revolutions
  by J. D. Bernal (Penguin Books, London 1969).
Science, A History by John Gribbin (Penguin Books, London 2002).
The Fabric of the Cosmos by Brian Greene
  (Penguin Books, London 2004).
The God Delusion by Richard Dawkins
  (Houghton Miflin Company, Boston 2006).
The Real Global Warming Disaster by Christopher Booker
  (Continuum, London 2009).
The Trouble with Physics by Lee Smolin
  (Penguin Books, London 2007).
The Undergrowth of Science by Walter Gratzer
  (Oxford University Press, 2000).

Index

A priori 82
Agricola 48
Aiken 66
Alexander 25, 31
Alfen 157, 163, 167
Alpher 175
Ampere 60, 128
Anaxagoras 21-23
Anaximander 18
Anaximenes 18
Animism 37
Anthropic Principle 172-179
Arche 17, 27
Aristarchus 45
Aristotle 8, 20-27, 31-35, 43, 49, 77, 99, 118
Artificial Intelligence 12, 64-68, 104
Astrology 31

Babbage 64
Bacon 44, 56-60
Bacteria 147-148
Barrow 174
Becoming and Being 20, 108
Bell 62
Bell Curve 182
Bergson 12, 21, 98, 106-111, 151-154, 168, 188
Berkeley 79-81
Berners-Lee 122
Bethe 175
Big Bang 119, 139-140, 157, 167-169
Black hole 130, 183
Blackbody 114

Bohm 99
Bohr 116, 131
Bosons 118, 121, 134
Bragg 50
Braho 46, 124
Branes 140-142
Broca 101
Bruno 47, 124
Burbridge 168
Byron 65, 80

Cabala 31, 39
Carnot 59
Carter 172-179
Casaubon 37
Casimir 133
Christianity 23, 31-34, 40, 45
Churchland 102-103
Cicero 23
Clausius 59
Complementarity 13, 98-99, 110, 111, 116, 117, 189
Cooke 61
Copernicus 45-46
Corresponence 37
Cosmic Microwave Radiation 164-165
Cosmological Principle 158
Coulomb 128
Crick 145
Crookes 53
Curvature of space 160-161

Dante 35

Dark energy 163
Dark matter 162
Darwin 21, 52-54, 92, 145-147
Davidson 101-102, 108
Dawkins 188
De Broglie 99, 116
Dee 44
Democritus 22-23, 49
Density 161
Descartes 8, 30, 44-49, 72-78, 92, 118, 128
Determinism 8, 54, 55, 185
Dimensions 120-121, 140, 173
Dionysus 19
Dirac 118, 131-135, 178
DNA 145, 153, 189
Doppler Effect 156, 159
Dualism 10, 28
Duration 106-107

Eccles 100-101, 110
Eddington 106, 177-178
Einstein 54, 99, 117, 129-131, 155-160, 163, 172, 181
Elan vital 12, 151-154
Electron 29, 53, 171
Emanation 33, 37
Empedocles 21, 27, 145
Empiricism 79-81
Enlightenment 9, 10
Epicurus 23
Epicycles 45
Euler 54, 127
Everett III 171
Evolution 17, 120, 146

Faraday 61-62
Faustus 44, 58
Fechner 22, 85-89, 97, 105-106
Fermions 118, 121
Ficino 36-39
Fisher 149
Flatness problem 165
Flud 44
Functionalism 28, 103-104

Galen 48
Galileo 44-48, 54, 127
Gamow 175-176
Gassendi 49

Gauge theory 136
Genes 145-146
Gilbert 40, 44, 128
Global Warming 186-187
Gluons 119, 184
Gödel 68, 105
Gould 147
Guth 166, 176

Hacking 173
Haldane 94-95, 149
Hamilton 54, 127
Hardy 9
Hatton 144
Hauy 49-52
Hawking 139.174
Heisenberg 115, 131
Heraclites 20-22, 109
Heresies 150, 167, 183
Hermes Trismagistus 9, 36-40
Higgs particle/field 7-8, 119, 137, 138, 141
Hippocrates 48
Hobbes 20
Holst 42
Horizon problem 165
Hoyle 149, 151, 157, 167-168, 176
Hubble 125, 156, 161
Human Genome Project 153
Hume 81-82, 188-189
Huxley 55
Hylozoism 9, 18

Idealism 28, 80
Inflation 166
Intelligent Design 143, 151

Jackson 102-103
James 92, 110
Jeans 178-179
Jeffries 38
Joule 59

Kant 33, 69, 82-86, 109, 129
Kelly 44
Kelvin 62
Kepler 46-48

Lagrange 54, 58, 78, 127
Lamarck 145

Index

Laplace 54, 58, 69
Large Hadron Collider 8, 121
Leibnitz 47, 64, 73-79, 90
Leptons 18, 30, 119
Leucippus 22, 49
Lobatchovsky 130
Locke 79-81, 107
Lodestone 29
Lötze 93
Lovejoy 24
Lovelace 65-68, 104
Lucas 104
Lucretius 23
Lyell 144

Mach 54, 89
Macrobius 35
Many Universes 141
Many Worlds 141, 171-172
Marconi 62
Marlow 43
Materialism 23, 54, 69, 102, 184-185
Maxwell 52-54, 62, 113, 128, 155
McDougal 90-91
McGrath 147
Melissus 20
Mendel 145
Mersenne 44
Mirandola 36-39
Monad 77
Monism 17
Moore 1223
Morgan 145
Morse 61

Nagel 105-106
Napoleon 60, 69, 87
Narliker 168
Neodarwinism 148
Neoplatonism 27, 32-37, 44, 47, 124
Neopythagorism 31, 46
Newton 30, 42-47, 50-57, 76, 113-115, 127-130, 159-162
Nietzsche 25, 40, 69-71
Nuclear synthesis 163-164
Numerals 63

Occam 54, 103, 172
Orphism 19, 25

Paley 143-144, 173
Panpsychism 22, 85-95, 106
Pantheism 11, 76
Paracelsus 48
Parmenides 20-23, 73
Pascal 57, 64
Penrose 139
Penzias 164
Perfect Cosmological Principle 158, 167
Peripatetics 25
Photons 115, 133
Physicalism 102
Planck 114-115, 131
Plato 8, 12, 20-27, 32-35, 68, 99, 109, 171
Plotinus 32-34, 45
Polkinghorne 147
Popper 100-101, 110
Porphyry 33
Potter 186
Pre-Socratics 8, 17, 27-30
Principle of Equivalence 156
Principle of Plenitude 24-26, 37, 171
Principle of Sufficient Reason 158, 169
Ptolomy 45
Pulman 186
Putnam 103
Pythagoras 12, 19, 22-27, 41

Quantum field 132, 137
Quantum theory 114-117, 123, 131
Quarks 18, 30, 119, 184
Quasi-Steady-State Theory 168

Reductionism 184-185
Relativity 54, 114, 117, 129-130, 155-160
Riemann 130
Röntgen 50
Royal Society 58
Russell 18, 110
Rutherford 50, 113

Salam 135
Saltation 147
Sayers 188
Schrödinger 88-90.117, 131-134, 172
Schwartschild 183

Searle 104
Shakespeare 44, 87
Sherrington 93-95
Sibelius 9
Socrates 23, 33
Soul 26, 28, 32
Spinoza 10, 73-79, 86, 111, 117
St. Augustine 35, 43
St. Thomas Aquinas 35, 43
Standard Model 119-124, 138, 185
Strawson 106
Strings 120-121
Substance 26, 74-75
Supersymmetry 139
Swift 58
Symmetry 51, 120-121, 135-139
Sympathies 37

Thales 18
Thodosus 181
Thomson 53
Thoth 36
Thrasymachus 24
Time 107
Tipler 174, 179-180
Tolkein 186
Torricelli 49

Transistors 123
Turing 67

Uniqueness 71, 173, 189

Vacuum 49, 128, 133
Voltaire 59

Watson 145
Weber 85
Weinberg 135
Wheatstone 60-61
Whitehead 14, 93-94
Wilson 164
WIMPS 162
Wittgenstein 92
Wordsworth 9, 12
World Soul 12, 19, 29

X-rays 50, 131

Yates 37

Zeno 20
Zero 63, 107
Zuse 66-67